78, book, 294

91, diff calc

29 R.A.A (Zeno)

The Physicist's World

2 15, NB ∞

32 N.B.

57 MATH

59 Aristotle a great systematizer. And def.

119 gravitational attraction; equal forces

295, Book on Newton

65, N.B.

78, Galileo

90, δ (delta) δ

91 "time interval"

As the time interval becomes shorter

91 Calculus history, 97, 99,

92 formulas for falling bodies, 93

90 symbol for average velocity

93 average acceleration

94 N.B convention. 100, N's 3 laws of motion.

Q: 100 (NB)

101 N.B mutual force of two bodies

force, 102, 105 (NB)

Uniform motion, 101

Wow!, 108

Whoops! 118

equation of motion, 117 $F = ma$

121 NB

122-3 NB ←

127

THE PHYSICIST'S WORLD

The Story of Motion and the Limits to Knowledge

2011

THOMAS GRISSOM

The Johns Hopkins University Press
Baltimore

Printed in the United States of America on acid-free paper
9 8 7 6 5 4 3 2 1

The Johns Hopkins University Press
2715 North Charles Street
Baltimore, Maryland 21218-4363
www.press.jhu.edu

Library of Congress Cataloging-in-Publication Data

Grissom, Thomas.
The physicist's world : the story of motion and the limits to knowledge / Thomas Grissom.
p. cm.
Includes bibliographical references and index.
ISBN-13: 978-1-4214-0083-9 (hardcover : alk. paper)
ISBN-10: 1-4214-0083-9 (hardcover : alk. paper)
ISBN-13: 978-1-4214-0084-6 (pbk : alk. paper)
ISBN-10: 1-4214-0084-7 (pbk : alk. paper)
1. Physics—Popular works. I. Title.
QC24.5.G75 2011
530—dc22 2010047477

A catalog record for this book is available from the British Library.

Special discounts are available for bulk purchases of this book. For more information, please contact Special Sales at 410-516-6936 or specialsales@press.jhu.edu.

The Johns Hopkins University Press uses environmentally friendly book materials, including recycled text paper that is composed of at least 30 percent post-consumer waste, whenever possible.

O Nature, and O soul of man! how far beyond all utterance are your linked analogies! not the smallest atom stirs or lives in matter, but has its cunning duplicate in mind.

Herman Melville, *Moby-Dick*

Contents

A Note to the Reader

The story we will share in the pages of this book, you as the reader and I as the author, contains a modicum of mathematics. I have used it sparingly, and judiciously, but to eliminate it altogether would have been dishonest, a form of intellectual deception and condescension, and an insult to your curiosity and intelligence. Telling the story of the physicist's world without the use of at least some mathematics is akin to trying to describe the experience of poetry to someone who has never seen or heard a poem. Physics, after all, is a mathematical description of the world. As you will learn, it is specifically a mathematical description of motion, of how all things in the universe move, from the tiniest to the very largest. Our story should convey by its form, as well as by its content, what it is about. We cannot properly comprehend how physics works without understanding something about the crucial role that mathematics plays in it; and the only way to accomplish that is through experiencing at least a small part of it for yourself. I have tried to promote that by taking some of the most important ideas and presenting them in a semi-quantitative way.

An important aspect of our story concerns the kind of limitations that physics in the twentieth century has imposed on our ultimate knowledge of the world. These limitations arise from efforts to describe the world quantitatively and can only be properly understood and appreciated in terms of those same quantitative ideas. To try to talk about them in any other way would unnecessarily obscure and mystify their significance.

I have not included mathematics where I thought I could do as well without it or where I thought that its difficulty would only confuse rather than clarify. I have included it in those cases where I regarded it as essential to properly understand the most fundamental ideas. I have restricted its use to those instances where the mathematics could be made readable rather than tedious and complicated. And in every case I have tried to say in words the same things that the equations express symbolically, in order that the one may reinforce and help explain the other, sometimes perhaps at the expense of repeating myself and becoming unduly redundant. I hope I have erred on the side of being too pedantic; I did not want to risk being unclear or misunderstood.

I would discourage you from skipping the mathematical equations, even on a first reading. They constitute an essential component of the story and should be considered even when you are not sure that you understand everything. Often what you find confusing will clear up as you continue if you first invest a little time in trying to read and think your way through the mathematical expressions. The key is not to omit anything but also not to become bogged down and stop just because some point is not absolutely clear. The struggle to understand is an essential element of the physicist's world just as much as is the exhilarating sense of elation and satisfaction that comes from finally succeeding. With only a modest struggle on your part, you will truly experience the drive and passion for understanding that from the very beginning has been the impetus behind the creation of our story. There is a deep pleasure in understanding these things, even if that pleasure is the only real objective.

Finally, this is not intended to be a history, and it should not be read, or judged, as history. It should be read more as an essay. It is a story, a narrative, and it is *our* story, not that of events or individuals in the past. It is not meant to represent the view of things at the time of the discoveries but rather to show how we use those discoveries *now* to make sense out of the world in our own time with the cumulative perspective and knowledge we have gained from all the discoveries of the past. It is a story about the shape that the physicist's world takes today and what that story tells us—and can never tell us—about the real world outside our window.

As with the telling of any story, there must be an eventual ending. This story is still ongoing, but I have purposely chosen to end it where I did. Omitted are the latest developments in particle physics and cosmology, as well as string theory (a concept, first proposed in the late 1960s, that views subatomic particles as one-dimensional oscillating strings, among other things). String theory to date is not really a theory at all but only an incomplete suggestion of how a theory encompassing all of the known forces of nature *might eventually* be formulated. In addition, what has been suggested thus far would produce a theory that in all likelihood could never be tested experimentally and thus might never be accepted as physics in the sense presented in this book.

None of these omitted developments alters the basic argument: the epistemology of twentieth-century physics has been one of *fundamental limits* on what it is possible to know about the world (and therefore attaining complete understanding is not a possibility). None of them alters that basic argument because they are all based on it; all of them incorporate the epistemological limits already discovered. The unanswered questions in these other areas are merely signposts along the path already marked by the discoveries of twentieth-century physics. Ironically, string theory, if it were ever realized, would constitute the ultimate epistemological limit, because it would be inherently unverifiable.

By way of background, this work comprises the intellectual outline and synthesizing structure of a two-quarter-long, full-time academic program that I taught by myself and with Neal Nelson, a computer scientist and mathematician, for several years at The Evergreen State College in Olympia, Washington. The intended audience was undergraduate liberal arts students who otherwise might not have taken any offerings from the regular science and math curriculum. The idea was to offer science more broadly across the curriculum and to give students some intellectually engaging and honest grounding in the nature and accomplishments of the physical sciences, without trying to form them into physicists. One measure of the success of the program was the number of students who were willing to tackle conceptually challenging material and the number who went on to pursue the natural sciences and philosophy, as well as intellectual history, more deeply in the curriculum and in subsequent graduate studies. During the twenty-two weeks of the program, they read this work in draft form, along with selections from the list of suggested readings and other related texts, which allowed them to delve more extensively into each of the topics presented here.

I want to thank especially my faculty colleague and teaching partner Neal Nelson for his unwavering enthusiasm and support for this narrative approach to presenting physics as the story of motion and the ultimate limits to our understanding of the world. Without his encouragement and prodding, this project might well have languished and gone unfinished. Thanks also to Lisa Jaggli of State Wide Printing in Albuquerque, New Mexico, for her skill and resourcefulness in helping prepare the manuscript and the figures. And my sincerest appreciation and gratitude to the editors and staff at the Johns Hopkins University Press—most especially Trevor Lipscombe, Greg Nicholl, Juliana McCarthy, and Michele Callaghan—for their reception and support of the manuscript and for their generous help and guidance in turning it into a book. And, finally, thanks to my wife, Rebecca, for her patient understanding and constant good humor during the long silences and many hours devoted to this book.

The Physicist's World

1

The Ancient Quarrel

With consciousness comes awareness, and with awareness comes storytelling. Of the different kinds of stories that we devise in order to explain things to ourselves and make sense out of the world, none is more fascinating or more central to who we are than that told by the physicist about the nature of the material world around us. The story of the physicist's world has its roots in the very origins of Western philosophy and has continuously evolved to reflect our changing understanding of the reality behind things. It began to take on its current form with the beginnings of modern science in the seventeenth century, and in our own time it has embraced the strange epistemology of twentieth-century physics. *Epistemology,* a word that takes us back to our Greek beginnings in Western thought: the inquiry into *what* we can know and *how* we can know it. What we can know limits the field of inquiry; how we can know it determines the nature of the inquiry itself.

I was still quite young when I first became aware of wanting to understand the nature of things, the *De rerum natura* of Lucretius, of wanting to understand everything—and to understand it completely—the way I already understood my immediate surroundings and the day-to-day events of my life. Of course, as I did, we all come to realize that this impulse arises out of an unconscious awareness that the world is mysterious, that even in the smallest and least significant of phenomena—like the burning of a candle or the splashing of raindrops on the parched earth—there is something unfathomable, something that drives us to seek refuge in the quest for understanding, even, if need be, in the *illusion* of understanding.

It is an impulse that led me to pursue physics when I might have done something else. My earliest interest was biology. I grew up out of doors and by inclination I was a naturalist, living out of field guides and identifying the birds, mammals, trees, and plants of my native soil—the Delta country of the Yazoo River basin in northwest Mississippi. Yet even then it seemed to me that the natural philosophers—the physicists—were onto a more fruitful approach. Out of a few simple principles came the impressive ordering of a vast array of phenomena, an ordering and an understanding that went far beyond the mere naming of things—things that seemed to my unschooled mind to share more differences than similarities. The diversity of biological phenomena appeared staggering and overwhelming. Newton's three laws of motion and the law of universal gravitation sufficed to order the entire universe. I found the illusion of this latter approach more appealing, and more satisfying, to my youthful craving for understanding.

With maturity and age, however, comes a sobering wisdom. During our lives we accumulate knowledge and a store of memories. But everything in our minds, and all our memories, are to some extent made up. We are the ones who remember. And either in the process of recording our impressions, or in recalling them later, we make subtle changes, molding and shaping what we know to agree with our cumulative experience of the world as we have lived it, each to some extent differently in our individual lives. There are matters of fact, to be sure: the sun rises and sets; we are born, live, and die. Beyond that, most of what we know is less distinct and is constantly being modified by each new thing that we learn and remember. This is as true of the physicist's world as it is the world of the poet or the philosopher, since all are creations of the mind, depictions of a reality that can never be absolute no matter how earnestly we seek it.

The absolute stops whenever we attempt to go beyond simple matters of fact, and facts by themselves are seldom of lasting interest. More important to us are the judgments we make about the plain facts of our existence, the world that we create in our minds to organize and explain our experiences—the stories that we tell ourselves about the world. The story of the physicist's world is but one more way of understanding the phenomena around us, and we shall see that it like all the others is very much a mental construct, distinct from the separate reality that lies behind things. The real world always is out there, beyond my window; and it remains forever illusive and unknowable, in ways that no depiction created by the mind can ever hope to overcome, no matter how strictly we may adhere to the rigid facts of our existence. We will find that in our mental creations even the facts take on a certain flexibility and irrelevance unsuspected before the twentieth century, though in retrospect perhaps it should not have been. We will examine the physicist's world for its limitations and shortcomings, as well as for its triumphs and truths. Both, we will discover, are equally interesting and equally revealing and both are essential to any complete understanding of what we can know about the material universe.

Our impulse to ultimately understand the nature of things found scant encouragement in the twentieth century. On the one hand, it was a century filled with remarkable scientific achievement, a time of true intellectual revolution, comparable to the beginnings of modern science itself in the seventeenth century. On the other hand, each singular advance has been built upon the discovery of insurmountable limitations in what we can know about the world. These limitations do not derive from imperfect knowledge or understanding or any practical considerations having to do with measurement precision or the like—in short, not due to human limitations—but result from the nature of the world itself. Nature, it seems, has built-in restrictions to what we are permitted to know, what it is even possible to know. Our instinctual suspicion that the world at its core is mysterious and unfathomable finds vindication in the epistemology of twentieth-century physics.

The twentieth century began with the theory of relativity and the discovery that the speed of light in free space represents an upper limit to how rapidly any form of matter or energy—and hence any information—can be transmitted. When we look up at the night sky, we see it not as it appears now, but the way it looked when the light first began its long journey to our eyes, a journey that even for the nearest star began more than three years earlier, and for stars outside our own galaxy began hundreds of thousands and millions and even billions of years ago. Any prospect of finding out what is happening now is limited by how far light can travel in one lifetime or in the relatively short span of human affairs. Our knowledge of everything beyond a tiny, immediate region of nearby space is confined to what occurred in the past. Most of the universe can be seen only in the remote past, and for us has no conceivably knowable future. We may view the macroscopic world, it seems, only by looking through a one-way glass that gives us glimpses of how it used to be. The farther we peer, the more remote in time the images that we see.

The quantum theory developed during the early part of the century has shown that we are likewise limited in trying to extend our knowledge in the other direction—into the remotely small and inaccessible regions of the universe. Here the restrictions are equally fundamental and are even more baffling and bizarre. Concepts of space and time derived from our direct sensory perceptions break down and give way to a loss of certainty in the meaning of position and speed, energy, and time. Strict determinism in predicting the outcome of an event is replaced by a set of vague probabilities designating the likelihood of each possible outcome, with the final result determined only by the act of measurement itself. The uncertainty principle restricts the precision with which the position and speed of a particle can be known simultaneously. The properties of matter become increasingly more numerous and complex and further removed from our direct sense perceptions as the nature and location of matter become more diffuse. We are permitted to know the position of a particle exactly only if we surrender all information about how fast and in what direction it is moving or measure its progress exactly only if we give up all knowledge of where it is located. The concrete realm of our senses devolves into a miasma of phantoms that steadfastly refuse to be pinned down. The best we can achieve is the uneasy compromise offered to us by the uncertainty principle, which can be taken as the fundamental axiom of microscopic nature.

Perhaps we should not be surprised by such limitations. After all, we are creatures of the intermediate, trapped somewhere between the twin remotenesses of the very small and the very large. Expressed in powers of ten, we are situated roughly midway between the size of the atomic nucleus and the distance light travels in a year, intermediate between the size of an atom and the size of a star, excluded from either extreme—by our infinitude, in the one case, and by our overwhelming bulk, in the other. Nothing in our direct experience can be expected to have prepared us for the nature of phenomena in either of these excluded regions. We cannot know what we have never experienced. Or at the very least our experiences in one realm cannot be expected to carry over into these other realms without surprises. Viewed this way, if we had been much larger we would have discovered relativity physics at the outset; or much smaller and we would have found quantum physics to be the correct depiction of events; and in either case Newtonian physics would seem like only a clumsy and crude approximation.

For a while it was possible to take solace in such a view. The physics of the microcosm and the macrocosm might seem strange to us, but everything was still all right in the world of our immediate sensory experiences. If I have never directly experienced an atom or moved at the speed of light, I can nevertheless feel secure in my knowledge of the world that I can actually see and touch. Yet even this solace has been short-lived. The discovery of deterministic chaos at the end of the century has identified physical phenomena directly accessible to our senses whose outcomes, though completely deterministic, are inherently unpredictable—and hence in that sense unknowable—even if all of the initial conditions are specified. We find ourselves banished from at least a portion of the last remaining stronghold of our most secure knowledge of the world. The irony of all these limitations is that the physicist's world gains, by its ability to examine the limits to our knowledge, what it loses in being unable to completely describe the real world.

We live in a time when it has become common to replace reality with some abstract creation of the human mind. In place of the real world we substitute a depiction of reality derived from a limited portion of our total experience. We *model* the world in some fashion. It is better—and more convincing—if the model is mathematical, and best of all if it is implemented on a computer. Even to those who should know better, a computer model somehow suggests that we have transcended the limitations of the human mind. We then use the model to tell us about the real world, which of course it cannot do or can do in only a very limited way. *Virtual reality* is just the most recent example of this process, describing as it does the latest fad of representing reality by computer-generated images, to which can be added other sensory experiences such as sound and smell and touch.

But virtual reality is only a new name for what has been happening since the rise of modern science in the seventeenth century. We have replaced history with theories of history—from Hegel to Marx to Toynbee to postmodernism. We have created political sciences, social sciences, political economies, and theories of psychology to model politics, society, economics, and the mind. All of these efforts stem in one way or another from well-intentioned but misguided attempts to mimic the startling success of Newtonian physics in understanding the workings of the material universe. The physicist's world is the prototypical virtual reality.

Yet understanding the simple motions of material bodies in the natural world is a far cry from understanding the infinitely more complex interactions of humans. Beyond the hope that springs eternal in the human makeup, the two have nothing in common. There is nothing in the success of physics to suggest that the same methods of inquiry (the *how we can know* of epistemology) will lead to any productive understanding in the vastly different and more complex arena of human affairs. Only a species more interested in ideas themselves than in what the ideas are about would be seduced into thinking it possible. For the success of physics is not due to the nature of the inquiry alone but to the field of inquiry as well.

The physicist's world is built on a very specific—and very limited—response to the twin questions of epistemology about what we can know and how we can know it. Its accomplishments are derived more from the restricted nature of the inquiry than from the methods employed. The success of physics is a very limited kind of success, one that can never be total or complete. The physicist's world is an impoverished and skeletal representation of the real world, in just the same way that those misguided disciplines built upon the shifting sands of imitation are an impoverished and skeletal misrepresentation of the real world of human affairs. This was true even before the startling discoveries of the twentieth century and to some extent in anticipation of those discoveries. With them, these limitations have become a permanent part of the very foundations of physical science, just as they are an intuitive and observable part of the real world. There are important lessons that we can learn from better understanding the mental world created by the physicist, from understanding what it can and cannot, or does and does not, tell us about the real world outside our window.

Part of the necessity for such lessons is that no one seems to have been listening to the real message of physical science in the twentieth century. Most of us gather our impressions of scientific progress from technology, not from the underlying science on which the technology is based. One does not need an understanding of quantum physics in order to be impressed with the workings of computers or television or satellite communications or CAT scanners. And because we are limited in what we can know does not mean that we are limited in what we can do or that the limitations in our actions will be immediately apparent to us. We see a world in which man has flown, walked on the moon, and sent spacecraft to explore the solar system and beyond. Our automobiles operate more reliably, last longer, and burn less fuel. With the flick of a switch, the lights come on. Computers have become faster, more powerful. The list is endless and can be tailored to fit anyone's particular experience or interests. Surely all of these impressive technological achievements cannot be based on a science with built-in limitations. Limitations are but temporary obstacles to be overcome. They represent nothing more, we believe, than the present imperfections in our understanding, not some fundamental property of the world itself. We are creatures who have looked into the face of God; by comparison, understanding the physical world should be child's play. We never stop to consider that it is really the other way around.

None of the epistemological limits inherent in the physicist's world is of any immediate practical concern in our everyday lives. Nor are any likely to become so in the near future or perhaps ever, for that matter. They are indicative rather of a certain attitude about the world. An attitude that lies at the heart of an ancient quarrel between philosophy and poetry referred to by Plato in his *Republic*. The poets, Socrates complains, tell lies about the nature of the gods, which, to Plato, meant telling lies about the nature of the world. Homer and the other poets told stories depicting the gods as capricious, arbitrary, quarrelsome, vindictive, spiteful, and vengeful. The world over which they had power was chaotic, uncertain, and unjust, subject to the whims of fortune and the workings of fate. Not even Zeus, according to the stories, could control fate. Some things were hopelessly and permanently beyond human understanding or influence, even, it seems, beyond the sanction of the gods themselves.

Such a world was anathema to Plato—and to philosophy. The aim of philosophy was understanding, and Plato took it as a given that the world was not beyond human understanding but was wholly amenable to reason and right thinking. Knowledge was perfectible, and the world could be made to yield its secrets to reason correctly applied. The gods themselves were reasonable and just; the true gods were in fact philosophers. Plato condemned the poets partly because he felt their view of the world was injurious to the establishment of a just and proper state. But he also wanted to believe with every fiber of his being that no question was beyond the reach of reason. The poets, for their part, recognized the misfortunes of life and its profound mystery as an essential and inescapable part of human existence. Aeschylus, Sophocles, and Euripides staged plays at the Dionysian festivals in fifth century BC Athens, holding up to their Athenian audiences questions without answers, upon which turned the inexplicable destinies of mortals. The Athenians were not dismayed or overwhelmed by the view of life presented in the tragedies. They found in them, as have audiences and readers ever since, a deep-seated reassurance derived from confronting and realizing the unfathomable nature of the world and from witnessing the essential dignity, courage, and heroism of mortals forced to act in the face of such daunting and overwhelming mystery.

Plato referred to this quarrel as already ancient. Likely, it has exercised the human intellect from the beginnings of consciousness and the use of reason. It has raged unabated since Plato first gave it permanent form in writing, with the philosophers laboring through the centuries to extend the realm of human understanding, and the poets pointing out to them its undeniable limitations. On merit the poets have always had the better of it. Their position in the quarrel is to defend the status quo. They have only to hold up the world clearly to our view and let each one forced to exist in such a world draw his own conclusion. In contrast, the philosophers—and here we would include the scientists and all who insist on viewing the world as rational—have had to invent worlds of their own—creations of the mind—with which to represent the real world, in hopes that we could be convinced that the world of their creation can in fact depict the world of our experience. The limited success of science has been taken as a sign of progress, and as long as there is evidence of progress the quarrel continues. Neither side can yet claim victory. Neither side will ever be able to. There is nothing in human existence to prove beyond any doubt that the world is unfathomable, nor can there ever be, short of a total failure of the imagination. And any failure on the part of philosophy can be viewed as but a temporary obstacle to be overcome by additional knowledge and understanding.

The scientific discoveries of the twentieth century have given encouragement to both sides in this ancient quarrel. Our understanding of the universe has become richer and more sophisticated by leaps and bounds.

Yet it is an understanding based on underlying constraints that clearly delineate the limits to our ultimate knowledge, in ways that seem unlikely ever to be surmounted. Indeed, assuming that these limitations are merely temporary is tantamount to acknowledging complete failure in our present understanding of the world.

The ancient quarrel is not just a trivial dispute over some philosophical nicety. It is, as Plato recognized, a profound disagreement between two fundamentally different world views: that of philosophy, which puts its faith in the triumph of reason over ignorance and the unknown; and that of the poets who believe that at some remove the world is, and will always remain, unknowable. The quarrel is waged across the interface between these two opposing views. What can be explained is not poetry, said Yeats, and there the matter comes to rest. It is an interface that runs unbroken throughout the story of the physicist's world.

Whether one's attitude toward the world is that of the philosopher or the poet is, however, not the main focus of our interest. Either party to the ancient quarrel would surely acknowledge the physicist's world as one of the greatest accomplishments of the human intellect. Those who view our existence as rational point to it as the ultimate example of what is possible, and the best reason to hope that philosophy may yet be vindicated. Those who hold a tragic view of our existence see its limitations and shortcomings as vindication of their own position. And very likely the matter will remain that way.

For the story of the physicist's world will never be finished. It is a story that in the twentieth century has revealed the constraints to its own ending and discovered about itself that it can never be completed. Not that it will ever end, for by its nature it is a story of unending detail. We will take it only as far as necessary to see in what direction it is headed, to see in what manner the rest of the story when it is written will most likely unfold. And there we will leave it, content not in the ending, for that can emerge only from the endless details, but assured at least about the shape of the ending. It is an ending to be written finally in the world that we see outside our window.

2

Motion

Motion, suggested Greek philosopher Heraclitus, is the defining characteristic of the world. Motion is what creates the world and brings it into being. It is the very essence of existence. Take away all other properties but that one, and the universe, or some universe, could continue. But without motion, nothing would be possible. Heraclitus appears to have understood the fundamental irony of our being: that the world can have permanence, or indeed any existence at all, only through movement and constant flux, making change the ultimate source of whatever constancy is possible. In the physicist's world, as in the real world, nothing is ever at rest.

We must be careful in this not to attribute too much to Heraclitus, although the indications are all there. His are merely the first faltering steps of which we have any record. All that survives across the 2,500 years separating us are some 124 scant fragments of writings attributed to him. They take the form of terse aphorisms marked by paradox and seeming contradiction. The style appears to have been a purposeful attempt to escape being too easily understood, and thereby misunderstood.

The fragments are like the world itself, puzzling, enigmatic, as if the truth about the world could be expressed only in the same contradictions that seem to lie behind reality. *Nature loves to hide*, he says in one place. He expresses much the same sentiment in other fragments: *The lord whose oracle is at Delphi neither speaks nor conceals, but gives signs. Unless you expect the unexpected you will never find the truth, for it is hard to discover and hard to attain. The hidden harmony is better than the obvious.*

Heraclitus seems to have demanded of his audience the same effort he had found necessary to arrive at his own understanding. *Let us not make arbitrary conjectures about the greatest matters*, he admonishes in one fragment. In others he says: *I have searched myself. It pertains to all men to know themselves and to be temperate. Men who love wisdom should acquaint themselves with a great many particulars. Most people do not take heed of the things they encounter, nor do they grasp them even when they have learned about them, although they think they do*. And the one that is perhaps most central to grasping Heraclitus' view of the world: *People do not understand how that which is at variance with itself agrees with itself*. For Heraclitus, it seems, understanding the universe means coming to grips with the essential contradictions apparent in all things and understanding them not as actual contradictions but as the essential nature of things. Echoes of this same warning about the difficulty of understanding permeate all of the fragments. It is their constant theme and the one thread that unites them.

Yet it is through the apparent contradictions of the fragments individually, and as a whole, that Heraclitus most clearly puts his finger on the essential nature of the world, which he expresses again and again: *Homer was wrong in saying, "Would that strife might perish from amongst the gods and men." For if that were to occur, then all things would cease to exist. It should be understood that war is the common condition, that strife is justice, and that all things come to pass through the compulsion of strife. Opposition brings discord. Out of discord comes the finest harmony. To be in agreement is to differ. The concordant is the discordant. Everything flows and nothing abides. Everything gives way and nothing stays fixed. You cannot step twice into the same river, for other waters and yet others go ever flowing on. Into the same rivers we step and do not step. The way up and the way down are one and the same. It is by changing that things find repose*.

The universe and everything about it is in a state of constant flux and change. All is motion and movement and change. Even the apparent contradictions are but changing and moving back and forth in our knowledge and understanding of the world. It is only through opposing views and conflicting ideas that any deeper understanding is possible; and the only real and lasting comprehension is that there is never the possibility of any absolute or unassailable understanding. The paradoxes are just that, merely apparent contradictions, not real ones. The seeming contradictions are fundamental, the truth, the nature of the world, the reality behind things. When we have discovered them and correctly understood them then we have understood the intrinsic nature of things.

Although he focused on the paradoxes, Heraclitus clearly believed in an underlying order, that the universe operated according to divine principles, or laws, which he termed the *logos*. *From out of all the many particulars comes oneness, and out of oneness come all the many particulars*, he says in one fragment. And more directly in another: *All things come to pass in accordance with this logos, although men are unable to understand it*. Even the *logos*, the fundamental ordering principle behind everything, is itself subject to the same paradox of being the understanding behind things and at the same time being beyond understanding.

Though we can only guess at the precise meaning Heraclitus intended, this particular fragment may well be the key to understanding all the others. It seems to express verbatim a theme running through the fragments as a whole: there is an order to the world, but it is intrinsically, and forever, beyond the reach of human understanding. This is not to be taken as a statement about the limits to human understanding but as a fundamental property of the world itself. The universe may have built into it inherent restrictions against any sort of absolute understanding. We will find this same notion directly embodied later in the epistemology of twentieth-century physics. The physicist's world has come to be expressed in terms that preclude absolute knowledge—and absolute understanding—just as Heraclitus suggested. Indeed this feature of the material world may be its most fundamental and most fascinating property.

The idea of a *logos*, a divine order behind things, was a Greek obsession, which should not be at all surprising, since a belief in an underlying order is common to any attempt to understand the nature of things. Understanding in the way we like to think of it—simply that of making sense out of things—begins with the assumption that there is some kind of underlying order to make sense of in the first place. There is no way of comprehending a world that is truly unordered or chaotic, in which events take place randomly for no apparent reason whatever and with no discernible pattern or connections. The only understanding possible in that case is one based on the principle that no understanding is possible, which, besides being contradictory, is contrary to our common experience of the world and is not at all what Heraclitus was suggesting.

Heraclitus does not see an underlying order that exists in spite of the seeming contradictions or an order that lies behind the paradoxes and explains them away. For him the paradoxes *are* the logos. They represent the only real understanding possible. The ideas expressed in the two fragments—*Everything flows and nothing abides* and *It is by changing that things find repose*—are not conflicting statements in need of reconciliation or further explanation. They represent the fundamental nature of reality, the logos itself, and are the key to understanding the world. This notion traceable to Heraclitus that our understanding of the reality behind things is governed by certain fundamental, intrinsic paradoxes is very much in harmony with attempts in the twentieth century to describe nature at the most elemental level, one that we will encounter specifically when we explore the quantum theory and the theory of relativity later in the book.

But it is primarily for his emphasis on the fundamental nature of change, or motion, that we are interested here in Heraclitus. It is for the saying: *You cannot step into the same river twice* that he is most often remembered. You cannot even step into it once, others would add, carrying the idea to its logical conclusion, and, thus, raising crucial questions about how to deal with processes that change continuously in time. These are questions that later thinkers had to confront as they constructed a consistent picture of change and motion in the physical world and to which we will turn our attention in future chapters. It is no exaggeration to say that motion is the defining characteristic of the world, that movement is quite literally what creates the universe, what brings it into being and gives it its existence. For all of the discernible properties of the world derive from motion.

Without motion, for example, there would be no time. If things did not move, then nothing would ever change, everything would remain always the same, and all of time would be compressed into a single, all-encompassing instant. There would be no past, no present, no future. Everything, and everywhere, in the universe would be at the same moment, and all parts of the universe would be in intimate and lasting contact. We would have no way of becoming aware of time, no reason to suspect it, no need for it. Our only inkling of time, and the only means we have of discerning and measuring it, comes from the observation of motion: from the swinging of a pendulum, the repetitive motion of a clock escapement, the oscillation of a spring, the vibration of a quartz crystal; the rising and setting of the sun, the nocturnal and diurnal movements of the stars, the progression of the seasons; the cycle of birth, life, and death, the beating of one's own heart, the cascade of thoughts through one's mind. Without motion time stands still. We may believe that we possess some intrinsic and intuitive awareness of time. That even as we sit in the stillness and isolation of an empty room, devoid of external sounds or stimuli, we are nevertheless able somehow to detect the passage of time. That feeling is an illusion, a learned response. It is the flow and progression of our own thoughts that we sense, the internal workings of our own mind and body, which too would cease along with any sense of time without the presence of motion.

Likewise, without motion, there would be no space and no concept of space. The universe would be compressed and united into a single, indistinguishable entity. If nothing could move, nothing would occupy space, as it is only by the movement of material bodies from one place to another that we are able to have—or ever need to have—the concept of matter taking up space or having size and shape. Without motion, all boundaries between material bodies and the surrounding space would disappear, as would the distinction itself between matter and space. It was by similar reasoning that Aristotle made space a property of matter and denied altogether the possibility of a void. To him the concept of empty space was meaningless, because space existed only by virtue of the matter occupying it. What Aristotle failed to recognize was the equal impossibility of bodies absolutely at rest, since the absence of all motion would likewise take away any means of differentiating between matter and space.

As we will discover before we are through, the only way we have of becoming aware of *any* of the properties of matter is through motion. Without it, matter and space would be completely indistinguishable, one and the same. The world would become uniform, homogeneous, featureless. Without motion, there would be no energy and no light, only a black, blank void. And without space or matter, the world would have no extent and no properties, and hence no existence. Everything that we observe is attributable to motion. The legendary "eppur si muove" (still it moves) supposedly muttered by Galileo at his trial and abjuration expresses not just a property of the earth and a principle of celestial mechanics but the creative act behind the existence of the entire universe.

Even the account of Genesis 1:2–3 (And the earth was without form, and void; and darkness was upon the face of the deep. And the spirit of God moved upon the face of the waters. And God said, Let there be light; and there was light.) traces the creation of the world as we know it to the act of imparting motion.

The question of existence has been one of the central concerns—some would say the only separate and legitimate concern—of philosophy from the beginning. Volumes have been written and entire lives have been lived in pursuit of the intransitive verb *esti, est, ist, is, to be*; as though existence is a separate, static property, distinct and distinguishable from any association with the action it gives rise to, or that gives rise to it. This separate pursuit of being is chimerical. For long ago, at the dawn of Western philosophy, Heraclitus suggested an answer to that question: there is no existence separate from action, separate from the creative act that gives rise to existence. The intransitive verb *to be*, by itself, conveys no meaning. Meaning resides only in the specific predicates of the verb, in the properties attributable to the subject, which are distinguishable solely by the actions they produce. The universe is born in action, through motion; it comes into being and is sustained only by ceaseless change. That essential insight into the nature of the world constitutes the true wisdom and most enduring legacy of Heraclitus.

Along with space and time, the notion of causality is fundamental to our experience of the material world. It may be argued that these three concepts—space, time, and causality—are not only necessary but sufficient for our purposes. We will leave that question to our story as it unfolds, for the concepts of space, time, and causality are key elements of the physicist's world. Our understanding of nature resides in whatever meaning and description we give to these three basic ideas. Not surprisingly, we will find that it is our understanding of these concepts that have been most seriously challenged by the developments in physics occurring in the twentieth century, in particular by the theory of relativity and the quantum theory. For the moment we want only to briefly consider what we understand by causality, in order to make a point about its rather obvious connection to motion.

When two events are associated in our experience, such that one is always immediately preceded or followed by the other, we adopt the notion that the first event causes the second. The first of these two events may in turn be thought of as having been caused by some other event immediately preceding it, and the second event may subsequently be the cause of yet a third event following it, and so on, in an unbroken sequence of occurrences all linked together in a continuous chain of cause and effect. One billiard ball strikes another, setting it in motion so that it in turn strikes a third, which then collides with a fourth, and so on. To claim a causal relationship between two events, one of them must always be observed to precede or follow the other in our experience, without fail. We cannot claim that event *A* causes event *B* unless we find that the occurrence of *A* leads to *B* every time. If *B* is observed to follow *A* only part of the time, then we are led to seek some other event that always precedes *B* as its immediate cause. And the closer together in time two events occur, the stronger or more direct is said to be the causal relationship between them.

Such vague notions of causality raise a host of difficult, even unresolvable, issues. For one thing simple causality rarely pertains. A single event may have a multitude of causes, all of which have to occur in some order to produce a particular final outcome. Identifying and sorting out such multiple causes may be difficult or impossible. There is the question of the time interval between events and how close together in time two events must occur for one to be considered the immediate cause of the other. In any scheme in which time is treated as continuous, any two events always have between them an unending succession of other events in a continuously unfolding chain of causality. There is also the deeper issue raised by David Hume in his skeptical attack on science and inductive reasoning, of whether two events can ever be linked together by any kind of logical necessity simply because they have always been observed to occur together in the past; or whether the possibility always exists that on the very next try the one might occur without the other. Causality, Hume argues, is not required as a matter of logical necessity but is merely a custom or habit derived from experience. Nothing in our experience can ever assure us beyond all doubt that a particular causal chain of events is a permanent and necessary feature of the world. Causality, Hume asserts, is rather an article of scientific faith. It was this latter assertion that stirred Immanuel Kant from what he described as his "dogmatic slumbers" and set him to arguing that Hume was surely mistaken—but that is getting ahead of our story.

For our present purposes, it suffices to point out that the whole notion of causality, like the concepts of space and time, derives from change and motion. A world without motion is without time and without events, without even the possibility of cause and effect, irrespective of whether we argue that causality is a logically necessary component of the natural world or merely a useful artifact of our present experience.

Clearly then, from all we have said, motion is the principle feature of the material world and the central problem of physics. It is in fact the *only* problem with which physics is concerned. James Clerk Maxwell described the dual concerns of physics as being matter and motion, but in actuality they are one and the same. We have to concern ourselves with the properties of matter only to the extent those properties are necessary to describe and understand motion. One consequence of this is that we can form no deeper understanding of matter than that based on our immediate sensory perceptions of space, time, and causality by which we understand motion. This is one of the inherent limits to our understanding that we will encounter as we explore the physicist's world.

We have no way of knowing how far or in what direction Heraclitus may have taken his ideas. The fragments are too incomplete to do more than suggest his thinking and the comments of later writers do not go beyond or add substantially to our understanding of the fragments. Certainly he would not have thought about these ideas in exactly the ways they have been presented here, where they are put in the context of the interests of our story, quite distinct from the chief concerns of Heraclitus. It is difficult though, from a careful and thoughtful reading of the fragments, not to draw some such conclusions or to imagine that they were at least part of what he had in mind. We know from commentaries by later writers, including Plato, Aristotle, and others, that Heraclitus's views came to be widely known and regarded. His ideas must certainly have been deemed significant and important. From the perceived obscurity of his writings, he acquired a reputation as a snob and a pessimist, neither of which are supported by a charitable reading of the fragments. He was apparently aristocratic and disdainful of the masses, by his own acknowledgement finding little wisdom in the common opinions of his day. And we know from the views expressed by some of his contemporaries that Heraclitus's ideas about the world were by no means universally accepted, and in at least one case were met by strenuous, almost vehement, opposition.

Heraclitus was but one of a number of early thinkers that today we lump together and refer to as the Presocratic philosophers. They lived during the late sixth and early fifth centuries BC, in Greek cities and colonies along the Aegean coast of Ionia and in southern Italy. Heraclitus was from the city of Ephesus in what is now Turkey. Only fragments of their writings survive for any of these Presocratic philosophers. Their designation as Presocratics places them chronologically as predecessors of Socrates, but more importantly acknowledges the debt that all philosophy, from Plato on, owes to the beginning made by this early group of thinkers. They were bent on nothing less than trying to understand the world. They asked fundamental questions about the nature of the universe, what it is made of, how it came into being, what governs its behavior. But the questions they addressed are not what distinguish them and give them their common bond. It would be the height of arrogance to assume that humans have not been asking these same questions since the dawn of consciousness. Questions like these are probably synonymous with human awareness.

What distinguishes the Presocratics are not the questions they asked but the criteria by which they judged acceptable answers. They rejected the old explanations, those attributing the nature of things to the gods, answers rooted in religion and in myth and stories. What Homer said was no longer good enough for them. They insisted, instead, on answers that met the test of reason—or rather reasonableness, for they were only beginning to probe the limits to reason. They had to be persuaded of the truth of any proposition and, in turn, sought to be persuasive in their own arguments. The Presocratics adopted the mind and reason as the final arbiters of truth and wisdom and banished the gods and replaced them with philosophy. This was an act of supreme confidence and optimism and one of intellectual daring and courage. Not only was the world understandable, it was within the grasp of human understanding. It also marked a pivotal point in our intellectual history, one, so far as we know, without precedence, confined as we are to what has been preserved in the written record. This moment marked the beginning of philosophy and the ascendancy of reason in human affairs and is one of the defining events of Western civilization. That would not be putting it too strongly.

The fundamental tenet of all this was that the world could be understood. That meant that it had to be based on some kind of underlying order. Since it was an order that could be comprehended by the mind, then that order must necessarily have the kind of properties that were consistent with reason. Hence reason itself, alone and unaided, should be capable of arriving at the underlying principles, as well as deciding the nature of the universe based on them.

Almost immediately then the Greeks began to use meta-principles, rules about the nature of the rules themselves. The use of reason is one such meta-principle. The concept that the universe is understandable is another. The view that the world should be beautiful, simple, pleasing, harmonious—because such concepts are somehow more in tune with how a well-ordered universe would be constituted—is yet another example. Such meta-principles became firmly embedded in Greek thought, and in many instances they survive unchanged in our own thinking. A more recent example of this is that a great deal of effort—so far to no avail—was expended in the twentieth century searching for evidence of something called "magnetic monopoles," because, among other things, their existence would complete the symmetry of Maxwell's equations describing the electromagnetic field. Completing symmetry is often a meta-principle for physicists. Discovery of these "monopoles" would have made Maxwell's theory even more mathematically elegant and render it that much more appealing to the human intellect. We will explore the work of Maxwell in chapter 10. The idea of symmetry and the concept of mathematical elegance are meta-principles of the human mind. Such ideas were at once the source of much of the success the Greeks enjoyed in asking the right questions and in posing tentative answers; but adherence to these ideas was also responsible for limiting the inquiry and shutting off other equally fruitful ways of answering. We shall encounter examples of both as we proceed.

Little wonder that these first efforts failed to achieve unanimity about the nature of the world. Reason, it seems, is not a very precise or exact instrument. Simple logic can be made rigorous and, for the most part, unambiguous. But even the most straightforward and precise form of reasoning—that of the syllogism based on the use of premises to arrive at a conclusion (First premise: If *A* is true, then *B* is true; Second premise:

A is true; Conclusion: *B* must be true.) can never be made totally certain and unambiguous. Slightly different premises can lead to completely different conclusions (if *A* is true, *B* is sometimes true; *A* is true; Therefore *B* may or may not be true). And the Presocratic philosophers possessed no universally agreed upon criteria for deciding the correct premises from among all of the possible choices, except for what each one could be persuaded was reasonable to accept. Reasonableness got mixed up with preconceived notions and meta-principles, and truth was variously attributed to one belief or another with no way of arbitrating between them. One can always be persuaded of what one is willing to believe; and where persuasion does not work all that is left is disagreement.

The Presocratics and the Greek philosophers in general had no tradition of basing their ideas about nature on empirical observations. They observed the world around them to be sure, and from their comments we find them to have been very astute observers. But they showed little inclination or willingness to accept the evidence of direct observations over the dictates of reason and belief. They mistrusted the senses and regarded them as suspect and unreliable. From the observation of such simple illusions as parallel lines appearing to converge in the distance and closer objects appearing bigger than much larger objects farther away, they were skeptical even about what they observed with their own eyes. Appearances were deceiving, and the senses in general could not be trusted to accurately depict what was real about the world.

Some of Heraclitus's skepticism and his preoccupation with paradox and contradiction may possibly stem from the perceived unreliability of sense impressions in conveying the true nature of the world. And reading all of the fragments imparts a strong impression of the undeniable importance that Heraclitus placed on reason. Yet there is no attempt on his part to shun the contradictions of the world, or to deny them. The fragments do not deny appearances but instead mirror them, in form and substance. The observable world is at least the starting point for everything Heraclitus had to say.

In the next chapter, we will learn about another group of Presocratics, whom we know today as the Eleatic philosophers, after the city of Elea in southern Italy where their founder Parmenides and his chief disciple Zeno lived. For them, reason alone—unaided by the senses—was considered a more reliable means of understanding reality, even when the conclusions arrived at through the use of reason were totally at odds with appearances. They vigorously attacked the views expressed by Heraclitus, using reason itself as their weapon. The result was a strange and starkly different conception of reality than the one suggested by Heraclitus.

3

To Be or Not to Be

Parmenides and the Eleatics seem in retrospect to have been greatly disturbed at the views expressed by Heraclitus. Their disagreement with his ideas about the nature of the world was so profound, and their response so radical, that encountering them today we are likely to find their position almost unfathomable, even pathological. Nevertheless it is undeniable testimony to their unshakable faith in the power of reason and the triumph of the mind over the senses. The dispute seems to have centered on the idea that motion is the principal feature of the world, that the universe is in a constant state of flux or change. The Eleatics were committed to understanding the world. But the notion that the world is in constant transition suggests that it may be beyond our understanding. By the time something can be known, it will have changed to something else, and the quest for understanding becomes the never-ending pursuit of an elusive and ephemeral quarry. Understanding, if it is to be possible at all, must be complete and absolute, for otherwise it isn't true understanding. Most of all whatever is understood must be permanent. Life in a changing world poses a threatening and uncertain existence; permanence alone is reassuring, both in the world around us and in our comprehension of it.

The most basic characteristic of the universe, thought the Eleatics, is not motion or change, but *existence*, which was a necessary prerequisite to everything else. Surely the most fundamental property of the world is not subject to change but is a priori permanent; otherwise the permanence of the universe itself can not be assured. As they looked around them, they saw evidence not for change but for the lack of change. The sun rose and set anew each day. The seasons came and went in the same unending cycle. The same stars kept returning to the same places in the night sky season after season, so regularly that men's lives and the conduct of human affairs could be ordered by them. Hesiod's *Works and Days* contained a catalogue of commonly held advice for planting and harvesting according to the positions of the various constellations in the sky at dawn and was based on centuries of observations. The permanent, immutable nature of existence as the fundamental feature of the universe was elevated by the Eleatics to the status of a meta-principle. Heraclitus with his muddled paradoxes and apparent contradictions seemed to be suggesting that no complete understanding was possible. He too took appearances as deceiving and taught that things were often the opposite of what they seemed. Better to reject appearances outright than to subscribe to an understanding based on the deception of appearances. And that is just what the Eleatics did.

Parmenides and his followers constructed an opposing view of reality in which motion, and any change at all, was impossible, a static, isotropic, homogeneous *oneness* whose fundamental, and only, feature was permanence. To do so, they argued from the nature of existence itself, using reason as their guide. The key to understanding the reality behind appearances, Parmenides taught, lies in making a correct distinction between the intransitive verb *is*, or to be, and the verb *becomes*. The choice is between *being* and *not-being*. Truth, said Parmenides, lies in the verb *is*; the verb becomes can never describe what is real since it implies a coming into being from a state of not-being. Only being, or existence, is conceivable. Not-being, by contrast, is impossible to conceive since it requires calling into mind—calling into being as it were—something that is nonexistent and hence incapable of being conceived. Thus only being is real. Not-being is inconceivable, and thus impossible, and hence could play no part in correctly representing the true nature of the reality behind things.

Never shall it be proven that not-being is, Parmenides argues in the known fragments of his work. *From that path of inquiry restrain your mind. The one way, that It Is and cannot not-be, is the way of credibility based on truth. The other way, that It Is Not and that not-being must be, cannot be grasped by the mind. For you cannot know not-being and cannot express it. It is necessary both to say and to think that being is. For to be is possible and not-to-be is impossible.* In these statements, note the explicit use of the mind rather than the senses as the final judge of what is real.

The idea of becoming, in particular, is an illusion since it depends on the possibility of not-being. For if one thing is to become something else, it must first go from being what it is to being what it is not, or go from being to not-being. But not-being is not real, because what is not cannot also at the same time be. Anything real could never be in a state of becoming since it would have to first not-be, or pass through a state of not-being, which is impossible. To the Eleatics the correct, and only, path to understanding is marked by the acceptance of is, or being, as true, and the rejection of is not, or not-being, as false. Everything else follows from that choice. *Necessarily therefore, either it simply Is or it simply Is Not*, Parmenides says. *Thus our decision must be made in these terms: Is or Is Not. Surely by now we agree that it is necessary to reject the unthinkable unsayable path as untrue and to affirm the alternate as the path of reality and truth.*

All change in such a world is clearly ruled out. Nothing can ever move since to do so it must first go from where it is (being) to where it is not (not-being). Nothing can ever become anything different from what it is. Permanence is the essence of being. The universe must be isotropic (the same in all directions) and homogeneous (the same throughout), since if it were not two adjacent regions would have to be different. But going from one region to the other would mean going from what is to what is not, an impossibility. By the same reasoning the universe must be spherical, since only for a sphere are all directions the same. We might likewise assume that it must also be infinite, since otherwise we would reach a place where the properties must change, where we must go from something to nothing (that which exists outside of the something). But nothing (no-thing, the state of not-being) is ruled out. Yet here Parmenides apparently reasoned differently, arguing that the universe must be finite. An infinite universe would be necessarily incomplete, and he rejected incompleteness (not-being-complete) in favor of completeness, in the same way that he rejected not-being in favor of being as representing what is real. Being, he said, is one, whole and complete, and hence limited or finite. The universe is without motion and unchanging and therefore has neither a beginning nor an end but has always existed and always will. Time then must be unlimited or infinite, and not complete.

A later Eleatic, Melissus, disagreed with Parmenides on this point, contending that the universe must be unlimited in both magnitude and duration, for the same reason in each case, which seems more consistent with the overall Eleatic stance. Melissus's argument against the possibility of motion is equally revealing of Eleatic reasoning. *There cannot be any emptiness*, he said, *for what is empty is nothing, and what is nothing cannot be. Accordingly What Is does not move; there is nowhere to which it can go, because everything is full. If there were some emptiness a thing could move into it, but since the empty does not exist there is nowhere for a thing to go*. This is only one of a number of arguments that the Eleatics offered against the possibility of motion. The Eleatics' rejection of the void became quite entrenched in the thinking of other Greek philosophers, including Plato and Aristotle, and was to persist as a popular doctrine well into the Middle Ages, and in the case of light and electromagnetism, as part of the physicist's world until the end of the nineteenth century. But as we shall see when we encounter the Atomists, this view, though prevalent, was by no means universal even among the Presocratics.

Finally, Parmenides reveals the true (and startling) extent of the Eleatics' commitment to reason when he says: *Thinking and the object of thought are the same. For you will not find thought apart from being, nor either of them apart from utterance. Indeed, there is not anything at all apart from being, because Fate has bound it together so as to be whole and immovable. Accordingly, all the usual notions that mortals accept and rely on as true—coming-to-be and perishing, being and not-being, change of place and variegated shades of color—these are nothing more than names. Thought and being are the same*. The world, Parmenides is saying, can only be known by thinking—through reason—and does not even exist separate from thought. The world of appearances—of change and motion and discernible differences—is reduced to merely a collection of empty names without substance or meaning.

Such an extreme and perverse view of reality as static, homogeneous, isotropic, featureless, devoid of any experience except what can be thought, and contradicting our sensory perceptions of the material world, if taken seriously seems enough to give philosophy a bad name. But that reaction would be missing the real point of the Eleatic position. Parmenides and his followers were not unaware of appearances. In addition to *The Way of Truth*, Parmenides also wrote about *The Way of Opinion*, the collection of views that men held about the world based on belief and the experience of the senses. Parmenides considered it necessary to be well informed about both ways. To live in the world, one must take into account the opinions of men. It is doubtful that the Eleatics any more than other men tried to walk through closed doors or that they insisted on standing in the way of large moving objects. What ultimate resolution, if any, Parmenides might have made between truth and opinion we can only guess, for the surviving fragments do not tell us. Nor do they even suggest any resolution. Quite likely Parmenides did not see any contradiction between the two different ways of viewing reality.

For Parmenides undoubtedly was trying to make a different point: that behind the world of appearances, in which we can be fooled and misled by the senses into believing the wrong thing, there must exist a deeper reality, one that is true in an absolute and unambiguous way, whose truth can only be approached through thought and reason thereby avoiding the deception of the senses. Otherwise we would have to abandon our hope of understanding the true nature of things. Parmenides clearly believed that the world of appearances was misleading and gave a false depiction of that deeper underlying reality and that Heraclitus was mistaken to see in the apparent contradictions the source of any kind of truth or wisdom. One way that the Eleatics dealt with the contradictions was to dismiss them as false and meaningless and to replace the world of appearances with a tidier more orderly world of the mind as depicting the true nature of reality.

The Eleatics took this step to its logical conclusion and made reality synonymous with mind. As Parmenides put it, *Thought and being are the same*. The world, that is the real world as opposed to the world of appearances, becomes literally the product of mind. Later thinkers would make the universe a thought in the mind of God. Thus God becomes all-knowing, but the truth is not revealed to us. The Presocratics insisted on knowing for themselves. The Greek gods already possessed human traits greatly magnified. In an act of supreme confidence in the ultimate power of reason, Parmenides elevates humans to the level of the gods and gives them access to true understanding. That has ever been the eternal hope and dream of the philosopher.

The specific conclusions reached by the Eleatics may seem strange and implausible to us, but the inclination and the desire to look beyond appearances for a deeper, underlying reality behind the nature of things should not. We find the world mysterious in ways that both thrill and frighten us. The mystery can fill us with an exhilarating and inspiring sense of wonder or leave us profoundly threatened and disturbed. We can deal with these feelings by celebrating the mystery and seeking wisdom in it as the poet does; or we can look for reassurance in understanding as the philosopher does, even if such understanding requires that we choose it over the apparent nature of the world around us. Others may find reassurance in the kind of understanding that stems from faith in religious or spiritual beliefs. Only the narrowest of distinctions separate these various kinds of responses, and the boundaries between poetry and philosophy and religion can become blurred beyond recognition.

Plato, and through him all later philosophy, was greatly influenced by Parmenides and the Eleatics. In the dialogues, Plato has Socrates meeting and conversing with both Parmenides and Zeno. The Platonic Forms—those idealizations and true embodiment of reality, of which everything in the material world is but an imperfect representation—clearly suggests the influence of Parmenides and the Eleatics. To Plato, the Forms are the ultimate reality. Knowledge of them is possible only through reason. True understanding of the Forms—discovering the real meaning of such concepts as Wisdom, Justice, Knowledge, Virtue—is the principle aim of philosophy. Though he tries to persuade us using reason, whenever he reaches the limits of such persuasion Plato is not above appealing to our desire to believe, and falls back on advocating philosophy as a kind of palliative faith in correct belief. At other times he seems almost to take a poet's delight in the seemingly insurmountable difficulties posed by the fundamental questions of philosophy.

Most of the world's great religions replace or supplement the apparent reality of things with some kind of belief in a deeper reality, a spiritual realm behind the material world. Many of us have no difficulty, and see little contradiction, in accepting such beliefs, without having to reject our more immediate experiences of the world. We are also perfectly capable of reconciling our faith in such beliefs with the view of reality constructed by modern science. Ironically, even science invokes a reality beyond the senses, beyond the realm of appearances. In probing the microcosm, the physicist explains observed phenomena in terms of entities like electrons, protons, neutrons, and quarks and concepts like mass and charge and intrinsic angular momentum, none of which can be seen or experienced firsthand without the mediation of an intervening apparatus. What we actually observe are the clicks of a counter, the needle of a meter, tracks on a photographic emulsion, or discharges in a spark chamber. These are at best indirect effects that can be attributed to the supposed unobservable entities. If challenged for more direct evidence of their actual existence, the physicist points to the mathematical models in which the entities and concepts in question play a direct role and have their origin. The mathematical models, in turn, will be defended by citing their internal consistency and the close agreement between predictions based on the models and the actual behavior of the measuring apparatus.

For the physicist, reality resides in the mathematical models by which nature can be described. The realm of direct experience is replaced by an underlying reality captured in the abstract symbols and formal relationships of mathematics. Like that of the Eleatics, the physicist's world too is a creation of the human intellect. And though guided in this case by empirical observations, the resulting depiction of reality is routinely extended far beyond the range of our senses, to explain the unobservable realms of the microscopic and the cosmic, the past and the future, as though the reality represented by the model is more fundamental than the limited observations on which it is based.

The Eleatic stance then should not strike us as altogether strange. It is not different in kind from what the mind does generally in trying to make sense of the world. Given the absence of any well-established tradition of empiricism, and the fascination of these early thinkers for the power of abstract reasoning, the Eleatic view of reality is a bold and imaginative creation. It is nothing less than an attempt to discover and understand the true nature of things by simply thinking about them and, not surprisingly, created a reality that was identical with thought itself. Encountering it today, we are apt to be struck by the inherent appeal that reason holds for us—the fascination and seduction of our own thoughts whatever their ultimate source—and with the difficulty of deciding what the concept of reality shall mean to us. We should not be surprised to discover that we are every bit as much the intellectual descendants of the Eleatics as we are of Heraclitus.

Zeno's role in all of this was to support the teachings of the master. He did so by trying to cast doubt on the positions taken by those who disagreed with Parmenides. Zeno made use of clever arguments devised to show that a reality based on appearances was even more absurd than the unity advocated by Parmenides. In the dialogue, *The Parmenides*, Plato shows Socrates playfully accusing Zeno of adding nothing new to the teachings of Parmenides, saying that the latter affirms unity while the former merely denies plurality; to which Zeno replies that the purpose of his writings is to support the argument of Parmenides against those who try to make him look foolish by deriving absurd consequences from his doctrine that all is one. Zeno's arguments are designed to turn the tables on those who believe in plurality by showing that on close examination their thesis involves even more absurd consequences than the doctrine of the One.

Only a few fragments of actual writings attributed to Zeno survive. One of these is an argument against plurality: *If things are many they must be finite in number. For they must be as many as they are, neither more or less; and if they are as many as they are, that means they are finite in number. On the other hand, if things are many they must be infinite in number. For there are always other things between any that exist, and between these there are always yet others. Thus things are infinite in number.*

Zeno employed the reductio ad absurdum form of argument and is credited by some with having originated it. In this form of reasoning, one begins by assuming the opposite of what one wishes to demonstrate and then shows that this assumption leads to some absurd, untenable conclusion, thereby demonstrating that the starting assumption must be false; and thus what one wishes to prove must be true. Literally, to lead (back) to an absurdity. This form of argument is a fundamental tool of logic and a cornerstone of mathematical proof.

In the fragment above one assumes that things are many in number. If they are many in number, then one can argue that they are finite. But if they are many in number, one can also argue that they are infinite. Clearly they cannot be both finite and infinite and hence we have been led to an absurdity (a contradiction). Since this absurdity is the result of the starting assumption (assuming of course that the rest of the argument is valid), then that initial assumption must be wrong. Hence things cannot be many in number and must be one.

A second example can be seen in another of the fragments, this one an argument against motion: *If anything is moving, it must be moving either in the place in which it is, or in a place in which it is not. However, it cannot move in the place in which it is (for the place in which it is at any moment is of the same size as itself and hence allows it no room in which to move), and it cannot move in the place in which it is not. Therefore movement is impossible.* Altogether a typical Eleatic argument.

Aristotle called this particular argument fallacious, saying that it depends on the assumption that time is made up of instantaneous moments, that is, that we can specify the exact position occupied by the moving body at any given instant. One supposes that Aristotle meant to say that motion is continuous and cannot be divided indefinitely into a set of discrete intervals of smaller and smaller duration, that there is no interval of time of zero duration corresponding to the instantaneous position of the body. If that was his meaning, Aristotle seems to be asking how a finite interval of time could ever be made up (even of an infinite number) of zero-duration intervals. And that is a fundamental question that goes to the heart of what we mean by the whole idea of continuity and limits.

It is a question that we can illustrate more clearly using another of Zeno's arguments, the one for which he is perhaps most famous. We know of this argument not from any surviving fragments of Zeno's writings but from others such as Aristotle who mentioned it in his own writings. For an object to move the distance between any two points, Zeno argued, it must first travel half the distance between them. Then it must travel half the remaining distance, and half of the distance remaining after that, and so on without end. No matter how many intervals the object has covered there will always be an interval left and hence the object can never reach its destination. Therefore motion is impossible. The arrow can never reach the target; the runner can never cross the finish line.

In another example usually attributed to Zeno, Achilles can never overtake the tortoise, since he must first cross the distance between himself and where the tortoise is now. But while he is doing that the tortoise will have moved an additional distance that Achilles will then have to cover, in which time the tortoise will have moved again, and so on, so that Achilles will always be left with a distance remaining between himself and the slower tortoise. The possible variants of this argument are endless, all designed to call into question the possibility of motion and to cast doubt on the world of appearances.

The puzzle posed by Zeno in this little conundrum is not trivial and is one that the Greeks were not prepared to deal with mathematically. We often see this argument referred to as *Zeno's paradox* because it involves a seeming contradiction, one that is only apparent and not real. The question embedded in Zeno's paradox, in whatever form it takes, is whether an infinite number of quantities, all of them non-zero, can be added together to give a finite result.

If we imagine motion the way it is presented in the paradox, the object must first traverse one-half of the total distance; then it must cover half the remaining distance, or one-fourth of the total distance; then half the distance remaining after that, or one-eighth of the total distance, and so on; so that at the end of each step the distance remaining is always half of what it was at the end of the previous step. The fraction of the total distance that has been covered after any number of steps can be written as the sum of the individual fractions traversed. At the end of the first step the distance traveled is $\frac{1}{2}$ the total distance. At the end of the second step the complete distance traveled is $\frac{1}{2}+\frac{1}{4}$ the total distance. At the end of the third step it is $\frac{1}{2}+\frac{1}{4}+\frac{1}{8}$ the total, and so on. In this way the fraction of the total distance covered can be expressed as the sum

$$\frac{1}{2}+\frac{1}{4}+\frac{1}{8}+\frac{1}{16}+\frac{1}{32}+\ldots$$

where the three dots signify that the sum never ends but contains an infinite number of terms (corresponding to the fact that there is always some remaining distance to travel), each of which is half the magnitude of the preceding term, just as Zeno claimed. For the object to cover the total distance between any two points, this infinite sum of fractions would have to add up to give a numerical value of 1. The question is how it could conceivably do that.

Although the terms in the sum become progressively smaller and smaller, there are an infinite (unlimited) number of them. In fact, no matter how many terms we include there are always an infinite number of remaining terms to be included in the sum. And each of them is greater than zero. How is it possible that we could ever add together an infinite number of values, each one greater than zero, and not get an infinite result? That is the fundamental issue raised by Zeno's paradox.

We can also look at the motion in terms of the total time required for the arrow to reach its target. Assume that the arrow is traveling at a constant speed and imagine the distance divided into intervals the way Zeno suggests. Since the distance covered in each successive interval decreases by a factor of two, the time required for the arrow to traverse each also decreases by a factor of two. The duration of successive intervals decreases but never becomes zero, and the number of intervals is infinite. Zeno wants us to conclude that the total time it would take for the arrow to reach the target must likewise be infinite.

Although a great deal has been written about Zeno's paradox there really is no fundamental difficulty here, at least not one regarding motion. We can easily demonstrate that the infinite sum of fractions above does indeed have a value of one. To do so, let x represent the unknown value of the sum and write

$$x = \frac{1}{2} + \frac{1}{4} + \frac{1}{8} + \frac{1}{16} + \frac{1}{32} + \ldots.$$

Then multiplying both sides of this expression by 2 yields

$$2x = 2\left(\frac{1}{2} + \frac{1}{4} + \frac{1}{8} + \frac{1}{16} + \frac{1}{32} + \ldots\right)$$

or

$$2x = 1 + \left(\frac{1}{2} + \frac{1}{4} + \frac{1}{8} + \frac{1}{16} + \frac{1}{32} + \ldots\right),$$

where in the second step we have explicitly carried out the multiplication indicated on the right side of the equation, term by term. But the expression in parentheses is just the original sum itself, or x, so that we can rewrite the last equation as

$$2x = 1 + x.$$

Finally, subtracting x from each side of this expression gives us

$$x = 1,$$

which is precisely what we set out to demonstrate. In the same way we can show that the time it takes for the arrow to complete its journey is also finite and is in fact just the result we would expect, namely, the distance traveled divided by the speed of the moving object. In either case—in terms of the distance traveled or the time required—there is no paradox.

As counterintuitive as it might seem at first, it is possible to add together an infinite number of quantities, each greater than zero, without getting an infinite result. The infinite sum of fractions above is clearly one example, but there are literally an infinite number of other examples, depending on how one specifies or chooses the individual values in the sum. Later we will see that, to deal with motion mathematically, we can break it down into smaller and smaller intervals such that the size of each interval approaches zero while the total number of intervals increases without limit, or becomes infinite. This procedure constitutes the basic method of the differential calculus and is one of the principal mathematical techniques by which the physicist's world is fashioned.

Although the Greeks were formidable mathematicians, their studies were primarily confined to geometry and observations about shapes. They dealt far less well with numbers. They possessed a working knowledge of arithmetic but they did not have even the most rudimentary form of algebra, which was a much later invention of the Arabs. The Greeks dealt almost exclusively with whole numbers and what are known as rational numbers, those such as $\frac{1}{2}$, $\frac{2}{3}$, and $\frac{3}{4}$ that can be represented as *ratios* of whole numbers. (Whole numbers are rational numbers, since any whole number can be written as the ratio of itself and one, e.g., $4 = \frac{4}{1}$.) They were familiar of course with the Pythagorean theorem, equating the square of the hypotenuse of a right triangle to the sum of the squares of the other two sides, and were the first mathematicians of which there is any written record to offer a rigorous proof of its validity. In fact they knew a number of different ways of proving this theorem one of which Plato presents in his dialogue *Meno*. They were equally familiar with certain irrational numbers like $\sqrt{2}$ and $\sqrt{3}$ that could be represented by the length of the hypotenuse or the sides of a right triangle using the

Pythagorean theorem, e.g., $\left(\sqrt{2}\right)^2 = 1^2 + 1^2$ and $2^2 = \left(\sqrt{3}\right)^2 + 1^2$. For a time they apparently thought that all numbers were rational, but we know that they eventually produced rigorous proofs that numbers like $\sqrt{2}$ could not be expressed as the ratio of whole numbers, hence their designation as "irrational."

The ancient Greeks could not have constructed a simple algebraic argument like the one we used to demonstrate that

$$1 = \frac{1}{2} + \frac{1}{4} + \frac{1}{8} + \frac{1}{16} + \frac{1}{32} + \ldots,$$

because they lacked the algebraic formalism on which it is based. But there is also a perfectly obvious geometric proof of this same result, obtained by noting that a line of unit length can be divided by repeated bisection into segments of length $\frac{1}{2}, \frac{1}{4}, \frac{1}{8}, \frac{1}{16}, \frac{1}{32}, \ldots,$ which when placed end to end make up the length of the original line. The ancient Greeks were unexcelled at geometry. Thus it is not likely that Zeno's arguments against motion would have been very compelling for long to anyone not already disposed to accept them. It is not even likely that he took them seriously. He was not trying to prove his argument but to cast doubt on the arguments of those who disagreed with Parmenides.

Whether the Eleatics convinced anyone else, or were even convinced themselves, that the world of appearances is really an illusion in which we are confused and deceived by our senses at every turn, is unimportant. They did succeed in focusing attention on the difficulties of using the senses as the ultimate measure of reality. Heraclitus suggested that motion is the chief characteristic of reality, that every attribute of the world derives from motion and change. The Eleatics suggested that behind the illusory realm of appearances lies another kind of reality, a unity and oneness, undifferentiated and unchanging, revealed only by reason and synonymous with thought. It is reassuring to think that behind the uncertain fortunes of this world, ever changing in ways that seem beyond our comprehension, there lies something permanent and unchanging that we can rely on and understand. Such a view must have seemed especially reassuring to these Presocratic thinkers, whose religion was based on a pantheon of powerful but capricious and willful gods embodying all the human follies and foibles. Reassuring, perhaps, but difficult to square with our experience of the world.

Where two ideas are so diametrically opposed neither is likely to endure for long in its pure form. The same faculty of reason that gave rise to each seeks some way out of contradicting itself. Still other Presocratic thinkers sought to combine these two conflicting concepts of reality, to synthesize a view of the world that incorporated both the plurality and change of Heraclitus and the unity and permanence of the Eleatics. In so doing they planted the seeds for a way of thinking about the material world that was eventually to become the foundation of the physicist's world.

4

Atoms and the Void

The starkly contrasting views of reality expressed by Heraclitus, on the one hand, and by Parmenides and the Eleatics, on the other, represent the two philosophical extremes of Presocratic philosophy. Still another group of Presocratic philosophers, whom we will refer to collectively as the Pluralists, sought in their various ways to affect some sort of synthesis of these two opposing positions. Unlike the Eleatics, they did not believe that the All was One but held instead that things were many or plural. Yet they sought a way to hold on to the unity behind things while being able to explain a world of diversity and change. Their aim was to save the appearances without having to give up on reason or the hope of understanding the reality behind appearances. In this they were thoroughly modern, which is to say they embraced a concept of reality that by and large we would still recognize as being in touch with our own.

Among the Pluralists were a number of individual schools of thought. Common to them all was that they looked for something fixed and unchanging, not singular but plural in number, out of which everything else was formed in combinations that could explain what we observe. The Pluralists accepted the Eleatic position that reality, however defined, is unchanging, but they rejected Parmenides's denial of the plurality revealed by our experiences. To them the material world is real. They adopted the basic argument of the Eleatics that later came to be expressed in Latin as the cosmological principle *ex nihilo nihil fit*, that out of nothing, nothing can be produced. Being, they argue, cannot arise from not-being. That which is in the world does not stem from what is not. Nor can being ever give way to not-being. Being, which represents the true reality behind appearances, is immutable and imperishable. But they did not accept the Eleatic doctrine of no change, or the arguments against change, since, as Heraclitus pointed out, motion is central to our experience of reality. In this they accepted the evidence of the senses, aided by reason, over the dictates of pure reason alone. They looked for an explanation of change in terms of an underlying unchanging reality.

Empedocles, for example, in surviving fragments of his writings, says: *They are fools, with no ability to reach out with their thoughts, who suppose that what formerly Was Not could come into being, or that What Is could perish and be utterly annihilated. From what utterly Is Not it is impossible for anything to come-to-be, and it is neither possible nor conceivable that What Is should utterly perish. For it will always be, no matter how it may be disposed of. And I shall tell you something more. There is no birth in mortal things, and no end in ruinous death. There is only mingling and interchange of parts, and it is this that we call nature.*

The specific parts that combine to form everything else Empedocles identified in another fragment: *How out of water, earth, air, and sun, mingled together, there arose the forms and colors of all mortal things by the unifying power of Aphrodite*. The four basic elements that account for everything that we observe in the physical world are earth, water, air, and fire. Note that these give rise to the material substances—forms—and their qualities—colors—that we observe. To Empedocles and other Greek thinkers, qualities such as color, temperature (hotness and coldness), and hardness are as ontologically real as material substances and have an independent existence on a par with substances. These thinkers possessed no concept of qualities as arising from the interaction of the material world with the senses but viewed them instead as separately existing properties or attributes of matter itself.

Note also that the uniting of the parts—the mingling together—came about through the influence of Aphrodite, signifying Love. Empedocles held that everything in the universe occurred through the action of two great opposing forces: Love and Strife. Love is the unifying or creative force; Strife is the dissipative or destructive force. The two wage constant warfare, the one prevailing at times then giving way to the other in an unending cosmic cycle. In this view, we find reminders of Heraclitus's emphasis on strife and harmony, the concordant and the discordant.

Here, though, instead of seeing them as the same, they become polar opposites out of which everything arises. The conflict between Love and Strife produces the specific changes that we observe in nature. At times there is harmony and creation, followed eventually by disruption and destruction, which in turn give way once more to harmony and rebirth. Other Greek thinkers, notably Plato, displayed this same preoccupation with polar opposites—good and evil, justice and injustice—always trying to find in the extremes some precision of meaning that would lead to understanding the true nature of each. Strife and Love, Empedocles believed, are also permanent and unchanging, like the four elements. When Love prevails, Strife withdraws but is not destroyed or changed, and vice versa. *Now one prevails, now the other*, he says in one fragment, *each in its appointed turn, as change goes incessantly on its course.* But, with respect to their perpetual cycle of change, they are unalterable and eternal. Plato later attributed to the Forms this same kind of permanence and made the Forms the unchanging reality behind the changing world of appearances.

Nowhere in the surviving fragments nor in any of the comments by other writers does Empedocles elaborate on exactly how the four physical elements were supposed to combine to give rise to all of the observable properties of the world. The only mechanism he proposes is the opposition between Love and Strife. Presumably to someone who lived in the world as it was then conceived, that suggestion was enough to make Empedocles's idea of the four elements believable.

His particular choice of elements was certainly not unreasonable. To anyone living in an agricultural society in intimate contact with the natural world, earth, air, and water are very visible and vital components of all existence. Grains grow from the combination of earth and water. Animals grow from eating the grain, drinking the water, and breathing the air. When things die, they decay and appear to turn back into earth and water and air (in the form of putrefaction and odors). Fire is a visible component of change, accompanying the transformation of material bodies into ash (earth) and rising air (smoke and fumes). Fire is hot, and heat is generated by the decay of vegetation into humus, or earth. The four fundamental elements span a wide range of densities. The most dense is earth from which come heavy metals like copper and iron. Water is less dense and is displaced by the heavier earth that lies below it. Air is lighter still and is displaced by both earth and water. The lightest element is fire, as can be seen from the fact that heated air rises and is displaced by each of the other three elements except when it is combined with them to form other substances. Combinations of these four elements in varying amounts could be used to explain the observed densities—and qualities like temperature and hardness and structural strength—of other substances, in a scheme that at least was reasonable to someone just beginning to try to understand the material composition of things. A more detailed mechanism or more elaborate explanation was not necessary to make such ideas plausible to Empedocles's contemporaries.

We know that Empedocles's ideas were widely regarded, if for no other reason than Aristotle spends so much time discussing them in his own writings. Aristotle's comments add substantially to our knowledge of Empedocles's views. Aristotle says that Empedocles took Love and Strife to be the source, or cause, of motion. When Love is at work unifying the world (or making one out of the many), or Strife is at work tearing down (making many out of the one), then there is motion. In the intervening periods things are at rest. Aristotle credits Empedocles as being the first to speak of good and evil as first principles, saying that Empedocles claimed Love is the cause of good things and Strife the cause of evil things. Aristotle criticized the view that all matter consists of only four elements, suggesting instead that things retain their separate identities when they combine to form other substances. This particular criticism goes to the heart of the principal weakness of Empedocles's pluralism.

In substances that are close to the pure form of the elements, it is easy enough to see evidence of the unchanged form of the element in the substance itself. It is reasonable for instance to believe that wine contains water unchanged or that iron heated white hot contains pure fire and earth in combination. And one might be convinced that molten iron displays the separate properties of earth, fire, and water in combination. But when the iron solidifies, no water is observed being given off, and yet the solid iron does not display the separate properties of water. It is no longer viscous like a liquid but solid. Since it must still contain the water, then the water must have somehow changed its form, casting doubt on the choice of water as one of the four unchanging elements that constitute the fixed reality behind appearances. It becomes difficult to imagine any small number of real substances, complete with all of their physical properties and observable qualities, out of which, unchanged, one can construct the astonishing diversity that we observe in the material world.

Another of the Pluralists, Anaxagoras, escaped this criticism with an entirely different approach to plurality. He did not propose to explain the enormous diversity of the world in terms of only four basic elements or any number of elements. For, as he put it in one of the surviving fragments of his writings, *How could hair come from what is not hair, or flesh from what is not flesh*? Likewise, we might ask, how could what is not earth come from earth or what is not water or not air or not fire come from water, air, or fire? Instead he proposed that *In everything there is a portion of everything else. Whatever there is most of in particular things determines the manifest nature that we ascribe to each*. In other words, everything in the world consists of a mixture of small bits—Anaxagoras termed them "seeds" or elements or parts—of everything else, all mixed together. The resulting mixture takes on the overall nature of whatever seeds it contains the most of. One substance is distinguished from another by the relative proportions of the ingredients in each.

Anaxagoras's views seem to have been biologically inspired, which is not surprising since the world of living organisms is one of constant mixing and generation and transformation. Blood resembles blood and bone resembles bone because one contains a preponderance of the seeds of blood and the other of bone. In the food we eat there must exist seeds of each, along with the seeds of all the other parts of the body that derive nourishment from that food. And in the organs of our body must likewise be found scattered the remaining seeds of the food we eat. Anaxagoras extended this scheme to cover not only things but qualities as well. Green could not come from what is not-green, nor hardness from what is not-hard, nor hot from what is not-hot. In this way of thinking about the material world no fundamental distinction is made between the nature of things and the nature of their attributes. Both are looked upon as common ingredients in the mixture that comprises the whole. The distinction between matter and its properties, between material substances and qualities—between noun and adjective—was not well-formed in the time of the Presocratics, nor was the distinction clear even among later thinkers like Plato and Aristotle, nor did it become so until after the development of modern science was well under way in the seventeenth century. And there was no general recognition of qualities as originating from the interaction of inanimate matter with the senses, even though as we shall see such ideas were very much a part of the Atomists' thinking. The qualitative pluralism of both Anaxagoras and Empedocles (who included fire, and by implication hot and cold, as one of the four basic elements of all substances) was very much in keeping with the prevalent views of the material world and its properties.

Anaxagoras went some distance in granting legitimacy to the senses, saying in one fragment that it is only *Because of the weakness of our senses* [that] *we are not able to judge the truth* and in another that *Appearances are a glimpse of the unseen*. The world of appearances is not some grand deception but only constitutes an incorrect and incomplete impression of reality because of the limited capabilities of our senses.

Since in any portion of a substance there is always a portion of everything else, then no matter how much of each of these ingredients we remove there must always be some quantity of each remaining. The doctrine of universal mixtures implies the concept of inexhaustibility. As a result there can be no least or indivisible quantity of any substance, because any quantity, no matter how small, must always be composed of quantities of everything else. In order for a thing of finite magnitude to exhibit inexhaustibility, Anaxagoras was forced to postulate the infinite divisibility of a substance. In this case, we have the same concept of infinitesimals—quantities that grow smaller and smaller without limit but which never reach zero magnitude though they approach arbitrarily close to it—that we find in the example of Zeno's paradox and that later on will become important in Isaac Newton's creation of the calculus. Anaxagoras was also obliged to conclude from the infinite divisibility of a substance that the quantity of elements or seeds in any substance must likewise be unlimited, and hence any object regardless of size contained the same number of parts. Here we have the germ of a mathematically valid concept of the infinite.

Anaxagoras made Mind, or *Nous*, the organizing principle, the logos, of his universe. He imagined a world that in the beginning was a completely random mixture of elements so that nothing could be distinguished from anything else. Initially the parts of the universe were all at rest. Since nothing could be distinguished, then in effect nothing existed. Anaxagoras seems to have argued that the universe was originally without distinction due to the unordered mixture of all its parts, but as we have noted before it would have been equally indistinguishable as a result of everything being at rest. Just as the human mind through the use of reason can impose order where none existed before, Anaxagoras postulated the existence of unlimited, autonomous Mind, pure and unmixed with anything else, that imposed on the undifferentiated universe the order that we now observe. Mind brought order to the universe by setting its parts into motion and by imparting to the entire universe a rotation that separated the lighter elements from the heavier in a kind of centrifugal principle described in the fragments: *And Mind took charge of the cosmic situation, so that the universe proceeded to rotate from the very beginning. When Mind first set things in motion, there began a process of separation in the moving mass; and as things were thus moving and separating, the process of separation was greatly increased by the rotary movement. The dense, the moist, the cold, and the dark came together here where the earth now is; while the rare, the warm, the dry, and the bright departed towards the farthest region of the aether.*

The one thing missing from this picture—though certainly present by implication—is any explicit mention of a gravitational principle. Otherwise it is strongly suggestive of our own notions about the formation of spiral galaxies, stars, and the solar system. In it rotational motion, and not simply translational motion in a straight line, plays a crucial role, just as the angular momentum associated with rotational motion came to play such a central role in the physical theories of the twentieth century.

Like Heraclitus, Anaxagoras considered motion to be the defining feature of the universe. Unlike Heraclitus, who believed that the universe had no beginning or end but always has been, is, and will be, Anaxagoras gives it both a beginning and a creator, making Mind the prime mover and the ultimate cause of all motion. If the universe exists because of motion, then something must have set it into motion. There is a clear distinction here between the two thinkers. To Heraclitus, motion needs no cause since it is the most fundamental property of the universe. Motion itself is the cause of everything, that which imparts to the universe its existence. Being in motion, not being at rest, is the normal state of matter in the world. Anaxagoras seems to have taken the position that, left to itself, matter would normally be at rest, a common view and one suggested to us by our observations of moving objects in the world around us. Things set in motion appear naturally to come to rest. A thrown ball falls to earth; a kicked ball slows down and eventually stops moving. Even an object sliding across a smooth slippery surface will only travel so far before it finally stops. It would seem that for motion to occur something must set objects normally at rest into motion and keep them in motion once they are moving; so the universe needed both a creator and an act of creation. For Anaxagoras, Mind supplies both.

Empedocles sees earth, air, fire, and water as the fundamental building blocks of matter, the unchanging reality behind appearances. Anaxagoras sees the infinitely divisible bits of everything as the fundamental constituents out of which the world is made. In his scheme, the number of different kinds of seeds is as many as the number of substances and properties they have to represent. The number of seeds of any particular kind is infinite, as is the extent of the universe and the amount of matter it contains. Both of these schemes were an attempt to identify what was constant in the world, in keeping with the Eleatic principle, while at the same time accounting for the apparent pluralism and diversity of the observable world.

We might imagine that Empedocles has proposed the simpler solution. Everything is made from only four elements. But the four elements must somehow account for the full range of all qualitative and quantitative properties that we observe for material substances, while keeping their own properties intact and unchanged when they combine with one another. Even though the properties of the four elements chosen are quite different, the observed diversity of the physical world is much too great for such a scheme to be convincing. It seems unlikely that simple combinations of any four substances whose properties are unchanged in the process could ever account for the vast range of material phenomena that we observe. Empedocles's scheme relaxes the Eleatic view of reality from a strict unity of one to a scant plurality of four, but in the process cannot convincingly account for how things appear.

Anaxagoras, by comparison, seems much more concerned with explaining appearances. The postulate that in everything there is a portion of everything else provides an immediate way of justifying the kind of material transformations that we observe in the interactions between substances. Earth and water and fertilizer combine to make the plant grow not because these materials are somehow changed into plant material, but because each of them already contains elemental bits of plant in themselves, which then come together to unite with the plant increasing its size. In the food we eat are already bits of blood and bone and flesh and hair and all the other distinct components of the body. When the body dies or the plant decays, these elemental pieces are returned to the earth and water and air from whence they came and from whence they are cycled again into other material forms. When Anaxagoras called the elemental units seeds, he no doubt had in mind the actual seeds from which we observe the plant growing and the human seed from which the embryo develops. He may have intended to say that everything contains within it the potentiality to be transformed into everything else, the way the seed has the latent potential for becoming the plant. If so he may have vaguely anticipated the idea of an atomic theory of matter. But he seems to have wanted to escape the vague notion of potentiality, which says nothing about the details of how things that are so different in their individual properties could combine to form something that is entirely different from any of its components. Instead he proposed, not the potential for becoming something else, but that everything in actuality contains portions of everything else.

This scheme directly supports appearances but at great damage to the unity and simplicity of the unchanging reality behind appearances. The number of elemental building blocks increases to include everything in the observable world that exhibits its own distinctive qualitative as well as quantitative properties. Just specifying all of the fundamental elements becomes virtually impossible and without hope of ever being complete and unambiguous. To accomplish universal mixing, the fundamental elements had to be infinitely divisible so that each one becomes infinitesimally small—and hence possesses no specifiable size—and the total number of them becomes infinite. Anaxagoras saved the appearances, perhaps only after a fashion, but to do so he had to relax the Eleatic view of reality from a strict unity to an overwhelming and unspecifiable plurality.

Another group of Pluralists—the Atomists—proposed a different kind of solution, one with features resembling in some ways each of these other two schemes but with an important difference: the Atomists replace the qualitative pluralism of Empedocles and Anaxagoras with a strictly quantitative pluralism. To do so, they have to abandon the most fundamental tenet of the Eleatic position: they grant reality to non-being as well as being. Like Heraclitus, the Atomists accept the reality of the observable world and make motion and change the principal feature and fundamental characteristic of the universe. They accept the Eleatic argument that without a void nothing would be able to move. But, since things clearly do move, then the void must exist. In what was a very uncharacteristic stance for Greek philosophy, the Atomists accepted the evidence of the senses, interpreted by reason, as having primacy over reason itself, or reason unaided. That is partly why we find their ideas about the nature of reality so familiar. But it is also because the general features of atomism end up being remarkably similar to the prevailing view of matter that emerges from physics and chemistry in the nineteenth and twentieth centuries.

Leucippus is the acknowledged founder of atomism, Democritus its chief proponent. Only a single fragment of Leucippus's writings survives, and it says nothing about atomism. From a number of fragments attributed to Democritus, we learn somewhat more, and we find out even more from the commentary of Aristotle and others. But by a remarkable coincidence we know more about atomism than about any other Presocratic doctrine.

The Athenian philosopher Epicurus, who founded a school in Athens and flourished in the fourth century BC, made atomism one of the foundations of his philosophy. Epicurus taught his followers that they should live so as to avoid pain and find enjoyment in life. One of the chief causes of unhappiness, Epicurus observed, is an irrational fear of dying and a morbid concern about the afterlife, promoted in his view by beliefs rooted in mythology and superstition and religion. If men but understand the true nature of the reality behind appearances, he believed, they would find nothing to fear in death. Epicurus taught that the truth about physical reality lay in atomism, which he took as demonstrating that after death nothing survives, not even the soul, that everything is dissipated as elemental atoms so that there can be no afterlife. We do not know how much if anything Epicurus may have added to the ideas of Leucippus and Democritus, but we do know that his teachings included a rich and detailed account of atomism.

This account was preserved for us much later by a remarkable Roman poet named Lucretius (Titus Lucretius Carus) in his long expository poem *De rerum natura* (On the Nature of Things) written around 50 BC during the social unrest of that period in Rome. Lucretius was a devoted follower of Epicurus and attributed much of what was wrong with Roman society, particularly the loss of optimism and concern about what the future would bring, to an unwarranted fear of death and to other false beliefs promulgated by superstition and religion. Like his master Epicurus, he believed that once men understand the true nature of things their fears will vanish. So he set out to enlighten them. And in the process he left us an altogether dazzling account of the atomic view of nature, one so complete and detailed, so in touch with our sensory perceptions of reality, and so full of evidence and reasoned argument that on encountering it today we are apt to wonder why it took almost two thousand years for us to arrive at much the same basic understanding of nature. That would be reading too much into Lucretius. The atomic view of reality that he espouses is by itself not especially convincing. It is a plausible scheme, but only one of many plausible ways of explaining what we observe. It is our own increased knowledge and greater familiarity with these same ideas that makes Lucretius seem persuasive. About the poet himself almost nothing is known beyond his name and some mention of him by others, notably Cicero. The poem seems to have had essentially no impact on his contemporaries and went virtually unnoticed. Only two copies of it survived, dating back to the ninth century, along with a single copy discovered in a monastery in the early fifteenth century. Yet it surely ranks as one of the most remarkable and amazing documents of all antiquity and presents us our most complete and detailed example of Presocratic efforts to think critically about the nature of the physical world.

In reality there are only atoms and the void, declared Democritus. Atoms are the smallest bits of matter into which any substance can be divided. Atoms themselves are not divisible, but being exceedingly small they are invisible. There are as many different kinds of atoms as there are basic substances in the world, and the number of each kind, like the universe itself, is infinite. Being indivisible, atoms possess size; they also possess shape and weight. These three quantitative attributes are their only specific properties and serve to differentiate each kind of atom from all the others. When atoms moving through the void come into contact, they can stick together to form various combinations or clusters, which can differ in their overall shape and in the ordering and spatial orientation of the individual atoms within the cluster. In this fashion all material substances in the universe are made up by various combinations of atoms.

The only certain way we have of knowing about the world, said the Atomists, is through the senses, which provide the only reliable source of knowledge and constitute the only source of truth. Only those theories that do not contradict the senses can be valid. The senses are to be interpreted by reason, as when parallel lines appear to converge in the distance but are known by reason not to; but whenever reason contradicts the senses it is reason that must give way, for to contradict the senses is to deny the true nature of reality itself. We must depart from the senses and resort to theory only where the limited power of the senses is unable to provide us with direct evidence, such as in the unseen realm of atoms. Lucretius finds plenty of indirect evidence for atoms though and cites a long list of examples: the slow thinning of a ring as it is worn over the years, the gradual wearing away of paving stones by the feet of passersby, the wearing down of an iron plow in the field; the way dripping water hollows out a rock, seaside cliffs eroded by the surf; clothes drying in the wind; the force of the gale that can whip up the sea, propel ships, and break great trees. In these cases and others, he says, we can observe the cumulative effects of large numbers of imperceptible bits of matter. Lucretius also cites evidence of voids in material bodies, pointing to such things as the seeping of water through stone and the flow of sap through wood to indicate that atoms are not tightly packed in matter. Whether we find his examples convincing or not, there is no clearer expression of empiricism and no firmer commitment to the use of empirical evidence than that which we see displayed throughout Lucretius's poem.

The number of different kinds of atoms must be finite since otherwise some of them would be large enough to be visible. The total number of atoms, however, must be infinite or else they would long since have accumulated in one place and we would not see matter distributed in every direction that we look. Since the number of atoms is infinite, the void must also be infinite or the infinite number of atoms would have filled it up. An atom is indivisible, and hence indestructible, since to destroy it would require breaking it down into something else. Likewise it cannot be created, since there are no parts out of which it can be made. Thus time, like the number of atoms and the void, must be infinite. The universe had no beginning and will have no end, but always was and always will be. The world that we observe was created entirely by chance as atoms moving through the void collided and clung together to form the various substances of which our world is composed. Not only was our world formed in this way, but many other worlds, corresponding to all the possible combinations of atoms, were likewise formed in other parts of the universe. The world is in a constant state of formation and dissolution under the continual barrage of atoms moving through the void. New material is constantly being generated as atoms collide and cling together in various shapes and arrangements. Existing substances are likewise dissipating as aggregates of atoms are broken up under the impulse of constantly bombarding atoms. These are all arguments employed by Lucretius in his exposition of atomism.

An atom can move under the influence of its own weight or as a result of colliding with other atoms. Initially atoms would presumably have all moved in straight lines in the same direction under the influence of their weights and would not have collided. According to Lucretius, Epicurus added to the ideas of Democritus and Leucippus the notion that at unpredictable times, and for no apparent reason, atoms can swerve away from straight line paths, thereby providing a mechanism for initiating a chain of random collisions that eventually gave rise to the order that we observe in the world. The Atomists incorporated this concept of *atomic swerve* as a permanent feature of matter and introduced an element of unpredictability in the motion of atoms, a loss of predictability that would have its counterpart in the quantum theory of the twentieth century in the form of the uncertainty principle.

One of the more fascinating parts of Lucretius's exposition is his account of the senses. All sensation is material in origin and is caused by the direct mechanical action of atoms impinging on the body, the senses, and the mind. Atoms striking the skin cause the sensation of feeling whenever we touch an object or the wind blows against our faces or we feel the heat from a fire. All objects give off likenesses of their shape and texture and even their color in the form of thin membranes, or "husks," of atoms from their outer surface that can cause images by impinging directly on the atoms in the mind or by striking the atoms in the eyes. Such atoms are very small and tenuous and can pass unimpeded through the interstitial spaces between the atoms of a substance. Dreams or images that come to us in our sleep or when we have our eyes closed are caused in this manner. Sound and speech consist of atoms too, argues Lucretius, as evidenced by the physical impact loud sounds like thunder have on objects they strike and the loss of weight suffered by the orator who exhausts himself speaking. When atoms of sound or speech strike the ears, they set the atoms of the body in motion causing us to hear and, in some cases to feel, the same sounds. Intense sounds are accompanied by such violent collisions that they can cause pain and damage the body.

In the same manner, taste and smell are produced when atoms emitted by certain substances strike the atoms of the tongue and the nose. In taste and smell, especially, the sizes and shapes of atoms play a major role. Pleasing tastes and odors are caused by smooth atoms that collide gently with the atoms of the body. Sharp pungent odors and tastes are due to atoms with sharp edges or prickly protrusions and those with spines or hooks that can grasp and tear as they make their way into the body. Substances that feel smooth and slippery like talc and olive oil consist of smooth round atoms that glide past one another with ease, whereas those that are rough or sticky have atoms that are barbed or hooked to grasp and cling.

On and on go the cases covered by Lucretius and the examples he uses to support the role of atoms. The result is an imaginative and on the surface most plausible and consistent attempt at formulating a completely mechanical model of all sensation. The result also is a totally quantitative pluralism as opposed to the qualitative pluralism of Empedocles and Anaxagoras. The only properties that need to be attributed to atoms are the physical properties of size, shape, and weight, each of which is quantitative in nature and intrinsically measurable. Qualitative properties such as color, hardness (or softness), heat (and cold) become atomic in nature, attributable to the quantitative properties of atoms such as their shape, arrangement, or motion. These are ideas that in slightly different language and with only slightly modified concepts are quite familiar to us today and hence sound somewhat more believable and persuasive than any actual merit behind them when first proposed by the Atomists.

The full range of phenomena that Lucretius invoked in support of the atomic theory in his poem is truly impressive: his goal is nothing less ambitious than to explain virtually everything in terms of atoms. Lightning, thunder, heat, and fire caused by friction, the working of the mind, plant and animal physiology, geology, astronomy, human society, law, ethics, history—all are dealt with as being understandable in terms of atoms and the void. In one place Lucretius even seems to be suggesting an atomic explanation for gravity. He notes that bodies are constantly being bombarded by atoms from all directions. Two objects in close proximity would shadow each other so that the number of atoms bombarding the facing sides would be reduced. There would then be more atoms striking the opposite sides, pushing the two bodies together. Interestingly enough, from purely geometrical considerations, such a force would vary roughly as the inverse square of the distance separating the two bodies, just as the gravitational force is observed to do, although Lucretius would not have known that. There is also a discussion of geometric optics that offers explanations concerning the operation of mirrors, the reversing of images by reflection, and the perception of distance by the human eye, all in terms of the motion of atoms.

Even the soul, which for Lucretius is synonymous with Anaxagoras's Mind, is postulated to consist of atoms that are infinitesimally small and delicate and highly mobile and that are easily influenced by those incident membranes of atoms that make up images. Not only can the soul detect images that are incident upon it, it can modify or transform them through the actions of its own atoms to form other images that were not there before, giving rise to thoughts and the sequence of thoughts that comprises thinking. At death, the delicate and tenuous soul is easily dissipated by the bombardment of other atoms, and its infinitesimally small atoms quickly escape from the body and are readily dispersed so that there is no afterlife. Nothing survives death. Everything is reduced once more to elemental atoms moving through the void.

The fear men have of death and the unknown, says Lucretius, is unfounded and irrational, the result of ignorance and religious falsehood and superstition. This is one of the constant themes of the poem. The other is the emphasis on the primacy of sensation as the one true means of acquiring knowledge about the world. The epistemology of atomism was both ambitious and optimistic. What we could know about the world was essentially everything; how we could know it was to accept the evidence of the senses and to interpret that evidence in terms of atoms and the void through the use of reason. This is basically the same view of epistemology that pervaded physics from the late seventeenth century to the end of the nineteenth century. To those unaware of the limitations of knowledge discovered in the twentieth century, it still seems like a reasonable view today.

But if we are to judge by the impact these ideas had on other thinkers then and later, as gauged by their comments and the amount of attention devoted to them, atomism was perhaps the least regarded of all the Presocratic doctrines. It was certainly the least regarded of the pluralistic solutions offered to resolve the conflict between the views of Heraclitus and those of the Eleatics. Plato takes no note of atomism and seems unaffected by it in his own thinking. He was generally disdainful of all the pluralistic schemes and was most influenced by the Eleatics in arriving at his own view of the Forms. Aristotle mentions the Atomists but mainly to show that he is aware of their views and to dismiss them as mistaken. He was especially critical of the idea of a void and made space entirely a property of matter in his own philosophy. But for having been included as part of the Epicurean philosophy, and for the fervor and passion of a Lucretius, we would know little of the richness and depth of the Atomists' thinking about the material world and the nature of reality.

Pluralism at least was an attempt to reconcile reason and the faculties of the human intellect with our sensory perceptions of the material world, and to do so in terms of the reality behind appearances—to identify what is real in the world, what *is*. Heraclitus and the Eleatics had succeeded in stating the problem; and the fundamental nature of the intellect is that any problem once identified must be addressed and pursued to wherever the final solution leads. The Pluralists acknowledged a world in which motion is real and change occurs, while still maintaining that something constant and unchanging must lie behind and beyond what can be known with the senses. Reason, mind, thought—consciousness—becomes inextricably intertwined with that underlying reality. That is understandable since it is through consciousness that we are able to know at all, and a fundamental tenet of any epistemology is that the knower becomes inseparable from what is known.

Heraclitus made motion the key to reality in what is still the crucial insight into the physicist's world. The position arrived at by the Atomists is not so very different from where Heraclitus began. All is atoms and the void—matter and motion. Matter has its own properties and these combined with the motion of atoms through the void give rise to everything that happens in the world, including mind and the human reason by which we make sense of it all. In the Atomists' scheme, the physical properties of matter are size, shape, and weight and are separate from motion. In the physicist's world, as we shall see, Heraclitus came closer to the truth: all the properties of matter are a consequence of motion. To see why, we must first learn how to understand the principles of motion.

5

Motion Constrained

The Presocratic philosophers could not arrive at a consensus concerning the nature of reality. They produced only differing views, with no way of choosing between them beyond the powers of persuasion to convince anyone what it was reasonable to accept. Reason was not then–nor is it now—capable of stripping men of their beliefs and preconceived notions. This was partly because of the nature of the questions being asked and partly because of the kind of answers that were proposed. It was also partly due to the absence of any long-standing tradition of empiricism among the ancient Greeks and their complete lack of any appreciation for the proper purview and limitations of reason. The use of reason in this regard would have to await a much better understanding of the types of questions and the nature of the possible answers. These first efforts were necessary, even essential, but they were also without the prospect of immediate success. They do however provide a valuable insight into what it was possible to achieve by simply combining reason with our common experience of the world, along with certain preconceived principles, and without the advent of later mathematical sophistication or any kind of systematic or quantitative empiricism to discover and test new hypotheses. Yet when measured against the goal of understanding the nature of things, the Presocratics were still far short of being successful.

The failure of these early efforts to yield unambiguous answers led to an overriding skepticism and pessimism about the possibility of a scientific philosophy. We find Socrates in the *Phaedo* confessing that as a young man he started out to study *phusis*—or physics, the natural science of the Presocratics—because he wished to know the true causes of things. He was especially taken with the view of Anaxagoras that Mind is what directs everything, because he believed that the directing Mind would naturally arrange each thing in the way that was best. To discover, then, the true causes of things, it would only be necessary to determine, by reason as it turns out, what is the best way for each thing to be and to act and be acted upon. Instead, he found that Anaxagoras and the other Presocratics looked for causes in the immediate interactions between material bodies. Socrates was not interested in immediate causes but in the ultimate causes of things, which could not, he believed, reside in the interactions of matter but must be the result of the best ordering of things, the design of some ultimate plan or purpose behind things, the logos of the Presocratics.

We note in this Socrates's (and Plato's) preoccupation with the world not as it exists, but as it *ought* to exist, along with all the preconceived ideas and judgments that accompany such a point of view. The distinction between the world as it is and as it ought to be is what separates science, or natural philosophy, from the rest of philosophy. We note also Socrates's (and Plato's) intention to let reason be the ultimate criterion by which we determine the best actions, which, in another guise, is merely the same thing as being concerned with the world as it ought to be rather than as it is. Socrates goes on to complain that, as a result of his early studies, he was led to doubt many of those things that before he had known quite clearly. It seems likely that these early studies and the failure of the Presocratics to reach any unanimous accord are what led Socrates to adopt his lifelong stance that he knew nothing and to devise the Socratic method of inquiry to convince others that they too possessed no sure knowledge.

Disenchanted with these early efforts, Plato turned philosophy away from natural science toward the affairs of men, toward moral philosophy with its concern about the nature of justice, right knowledge, and right behavior, which he called the Good. As a result, Plato contributed nothing to the discussion about the material world and essentially nothing to later attempts to arrive at a more satisfactory understanding of physical reality. In his surviving writings, Plato makes but little mention of the Presocratic philosophers. Some are occasionally mentioned briefly; others are omitted entirely. He seems to have been little interested in their ideas about the actual world and even less in the natural world itself. When he does discuss his ideas about cosmology, as in the *Phaedo* and the *Timaeus*, they are fanciful and highly speculative, mythological, and mystical, intended to invoke in the reader a symbolic and imagerial response rather than a literal or scientific one. Still, Plato is interested in metaphysics, in the ultimate reality behind things, and in ontology, the nature of being. But for Plato the answer to these questions resides in the Forms.

Some have wanted to view the Platonic Forms as representing merely definitions, the universal characteristics of physical objects like a chair or a stool, and of concepts like truth, knowledge, and justice—or as the correct understanding of what absolute and unequivocal meaning such terms could have. Viewed that way the Forms consist of nothing more than a collection of ideas or thoughts. But to Plato the Forms are much more than that. They are not simply definitions. In fact they are not definitions at all, since one thing can only be defined in terms of something else that then must be defined in terms of something else again, in what is a never ending and unresolvable regression. They are not merely ideas either, but are real, that is, have a real and actual existence. Not only are the Forms real, they are the *only* real things. Everything else in the world is less real since everything else is but an imperfect imitation of the Forms. There are many physical chairs in the world, but each is only an imitation of the ultimate Form of chair, which alone embodies the complete and absolute essence of all that is meant or can be conveyed by the designation *chair*.

Objects in the material world, even physical objects like a chair, owe their existence somehow to what Plato calls *participation in the Forms*, in this case the Form of chair. Exactly what he means by participation in the Forms is left purposely vague, though the connection between the Forms and the material world has something to do with Plato's concept of the soul.

Thus for Plato concepts like Justice, Truth, Virtue, Knowledge, the Good are not just words but are real entities with an absolute meaning, an absolute existence. To him, knowledge of the Forms is the one purpose of philosophy, and such knowledge could come only through philosophy. With that knowledge, we would have no need of natural science since we would understand the ultimate reality behind everything. No matter that Plato never quite got around to telling us how we are to go about acquiring knowledge of the Forms or even for that matter convincing us that it is at all possible. The means at least was clear: the right application of reason. The senses play a role, but only peripherally, mostly by providing us with specific but only imitative examples to guide our reasoning. The reality of the Forms is closely reminiscent of Parmenides's view that thinking and the object of thought are one and the same and that thought and being are the same. Plato does not dismiss the world of appearances as an illusion, but he sees it as only a pale imitation of reality. To him, the real world consisting of the Forms is not physical but mental, suggesting the extent to which he was influenced by the Eleatic school.

There is an interesting and related issue here concerning whether a certain kind of knowledge about the world is discovered or whether it is created by the human mind. Clearly there are certain matters of fact that are discoverable, things like the rising and setting of the sun, the motion of the stars in the night sky, the phases of the moon, the fact that material objects fall toward the earth, things of that nature. Such matters of fact represent a kind of knowledge about the world that is determined by experience. Hume, as we have seen, insisted that matters of fact could only be discovered through experience and could not be shown to be necessarily true. That is, we could not be assured that a given fact established by past experience would continue to be true of our future experience. There is, Hume argued, no *logically necessary* connection between such matters of fact. He means by that there is no logically necessary connection that can be established through experience alone. Nevertheless, we usually do assume that under the identical set of circumstances the outcome in any given situation will be the same the next time as we have observed it to be in the past. Our experience of the world, Hume said, conditions us to expect that but does not assure it.

But what about the kind of knowledge represented by a mathematical theorem, say, the Pythagorean theorem, $a^2 + b^2 = c^2$, relating the lengths of the sides of a right triangle, with c being the length of the hypotenuse, or the side opposite the right angle? Is the Pythagorean theorem a discovery about the world or a creation of the human mind? There is a logical connection between the concept of a right triangle and the Pythagorean theorem. One follows logically, and of necessity, from the other. For all right triangles, the square of the hypotenuse is equal to the sum of the squares of the other two sides. Conversely, any triangle for which the Pythagorean theorem is true must be a right triangle. But is the same thing true of the real world? Is a right triangle an actual thing or only a mental construct? For Plato, any actual physical triangle we can construct is only an imitation of the ideal right triangle (the Form of a right triangle), and for these imitations the Pythagorean theorem is only approximately, but not ever exactly, true. With greater and greater precision we can make closer and closer copies of an actual right triangle for which the theorem becomes more and more exact. But can there ever be a real right triangle? There would first have to be straight lines. But are straight lines even possible in the real (material) world, not just as a matter of greater and greater precision but even in principle? Perhaps a straight line in the Euclidean geometry sense of right triangles does not even exist in the material world. The theory of general relativity, for example, suggests that all lines in nature have curvature.

We might turn the process around and use the Pythagorean theorem as the definition of a right triangle. A right triangle is that figure that satisfies the Pythagorean theorem. Viewed this way a point, a line, a plane, and all of the other fundamental concepts of geometry are defined to be those entities that obey all the theorems of Euclidean geometry. Whether such primitive concepts as that of a point, a line, or a plane correspond to anything in the material world or merely represent concepts that are logically consistent with the axioms, postulates, and theorems of Euclidean geometry is left an open question. The existence of real points, lines, and planes become questions to be answered by physics, not mathematics.

Yet the remarkable success of mathematics in describing the physical world leads us to wonder whether it represents some kind of deeper truth, or reality, behind appearances. A great deal of mathematics has come about directly as a result of efforts to represent physical phenomena or to solve problems suggested to us by our experience of the world; so perhaps we should not be too surprised that mathematical formalisms seem to closely model the physical world. Even if one wants to argue that mathematics is a creation of the mind and not a discovery of some truth about the world, our minds are themselves products of the physical universe and may perhaps only function in ways that make even our most abstract constructions mirror the physical processes that gave rise to them. Perhaps we are incapable of mathematical formalisms that have nothing at all to do with the real world, only we do not always immediately see the connections. Quite often an abstract mathematical construction devised for one purpose—or purely for its own sake—will later turn out to be exactly what is needed for some entirely different and seemingly unrelated application involving physical phenomena. An example is the non-Euclidean geometries developed in the nineteenth century. They were motivated primarily by the desire to better understand certain concepts and postulates of Euclidean geometry, without any regard for immediate applications. But they led eventually to the very ideas and mathematical formalisms that half century later Albert Einstein would use to formulate his general theory of relativity, which revolutionized our understanding of gravity and the overall geometry of the universe. We will take up that part of our story in chapter 13.

Plato considered mathematics to be true knowledge. Further, he took it as the model for all knowledge. "Let no one ignorant of geometry [mathematics] enter here," read the inscription over the entrance to Plato's Academy in Athens. Without first understanding the nature of mathematics, one could not hope to understand the Forms and hence philosophy itself. He would have thought of the Pythagorean theorem and all mathematics as a discovery about the one true nature of the real world. The right triangle is real and is one of the Forms, in fact the model for all the Forms. Plato replaced the reality represented by the Eleatic unity with a reality based on the kind of knowledge represented by mathematics. The real world is abstract in the way mathematics is abstract; the physical world is but a material imitation of the true mathematical reality behind appearances. Plato's unity behind things is represented by the basic nature of the Forms themselves, a nature that is best illustrated in the abstract concepts of mathematics. At least that is about as close as Plato came to explaining it for us. It is to Plato that we ultimately owe our concept of a mathematical model, the idea that we can replace the physical world with the kind of mathematical representation, or *description*, on which the physicist's world is based. Only, to Plato, the mathematical representation would have been the real world, and the material world but an imitation of that reality.

Aristotle eventually disagreed with his famous teacher about the Forms. He did not regard them as independently existing real entities in the manner that he claimed Plato did. To Aristotle, the Forms are generalizations of certain universal properties or features exhibited by all members of a class of things; but they are *nomina* (names) and not *res* (things). They do not exist independently of the things of this world but are inseparable from them. Form could be distinguished from matter, but to exist as a real substance an object must possess both. The Platonic Forms are merely abstractions, and Aristotle was much more interested in the concrete world of specific things.

The type of abstraction that occupied Aristotle was that of looking for order in the world. He believed in a logos that governed the universe, and the logos was apparent in the order that existed in the universe, in the way it was organized and in the harmony by which it functioned. Things always have an explanation, and that explanation could be discovered by using reason to interpret our experience of specific phenomena. Aristotle was a great systematizer and organizer. As he thought about the world, he looked for ways of organizing his thoughts. The patterns of organization that emerged in his thinking became the order that he sought, and saw, in the world. He seems to have taken seriously Anaxagoras's view that Mind is what had imposed order on the world; hence it is by Mind that order could be recognized and understood. It is arguable whether the patterns so discovered reflect the true order of the universe or only one imposed by the mind of the observer.

Typical was the way Aristotle characterized the structure of a proper definition. First, the definition must place the thing being defined in a group or class of things with which it shares general characteristics, and, second, it must specify in what way the thing differs from the other members of the class. Genus and species, generic and specific: these concepts are as permanently ingrained in our way of thinking about the world as they are arbitrary and subjective in their application. Aristotle's example was the definition of man as a rational animal. Man is first of all an animal, but differs specifically from other animals by being rational. No matter that there are virtually limitless ways of formulating a definition of man that will fit into the same structure. Aristotle was not interested in uniqueness but in order. We see in this approach to definition no lofty preoccupation with Platonic Forms and absolute meaning, but a more practical concern with useful and usable ways of thinking about the material world that was typical of Aristotle.

Aristotle grounded reality—and his philosophy—in the physical world of appearances. He began by specifying the different kinds of realities that we can encounter in our experience of the material world. These are the famous Categories: substance (the primary one which each of the others presuppose, consisting of matter and form), quantity, quality, relation, place, time, position, state, action, and affection. But, whereas Heraclitus focuses on motion, Aristotle begins by making substance the primary reality and motion a property dependent on substance through the categories of place, time, and position. To him, substance is not the same as matter but is matter combined with form. Without form, we do not have a substance but only the potentiality for substance, which formless matter represents. Matter, in contrast, usually takes the form of some substance, such as that of wood or iron or stone, though matter always possesses the potential for taking on other forms as well, as when one substance is transformed into another. Thus Aristotle does not endow form with a separate and independent existence as Plato advocated but makes it a property inseparable from the specific material substances of the physical world. To him, form is not even a universal but varies from substance to substance, as in the specific differences between two objects both of which are chairs but that differ in design and materials of construction.

When we look at the world predisposed to see order, we find it, even if it may only be an order that reflects the operation of the human mind and the organization and patterns of our own thinking. And, having found order, it is only natural to see it as evidence of some purposeful arrangement of things according to an overall design or plan. Where there is a plan, it is natural to impute intent. And with intent we come full circle to seeing the world not as it is but as it ought to be or is trying to become, usually guided by our own preconceived notions or meta-principles. Aristotle viewed the workings of the world, and all natural phenomena, in terms of things happening to achieve some ultimate purpose. This was an attitude that characterized Plato's philosophy and about which Aristotle was in complete agreement with his teacher. Plato argued in the *Phaedo* that true explanation must always be *teleological*, in terms of achieving some end or purpose (from *telos*, or end). The Presocratic philosophers were wrong, he said, in attributing the motion of bodies to material causes, as in saying that one body causes another to move by acting upon it in some way. The material means tells only how the body is able to move, what enables it to be set into motion. The motion itself is for the purpose of attaining some intended end result, and that purpose is the true explanation of why the body moves. In addition, for Plato the end result invariably had to be one that was desirable or it could not be consistent with an ordered universe.

Aristotle too was interested in why things happen, but he was more interested than Plato in the details of events and the actual mechanism of how things happen. He expanded Plato's notion of causes to cover four types of causes. These are what Aristotle called the material, efficient, formal, and final causes, the four reasons that a particular thing has come to be this particular thing and not some other. The *material cause* is just the substance out of which a thing is made. The *efficient cause* is the already existing thing or being—the agent—whose action is necessary for the thing to come into being. The *formal cause* is the form that the thing or object takes, that which gives it its specific being and makes it this thing and not some other. The *final cause* is the ultimate end or purpose for which the thing comes into being. Of these four, we would recognize only the efficient and final causes as being somewhat like what we usually mean when we speak of the cause of something. The other two—the material and the formal causes—belong more to the category of how a thing comes to be: it is made out a material substance arranged according to a certain form. Plato intended his philosophy to be directed toward a knowledge of final causes, which resided ultimately in the Forms. Aristotle held that final causes are grounded in the physical world of experience and cannot be separated from the material, efficient, and formal causes. The final and formal causes could be thought of as related, since the end or purpose of a thing coming into being was to achieve its desired form as well as possible, to be the most perfect example of a tree or a chair or a man that conditions permit. This was true in particular of natural objects, as distinct from manufactured ones.

Like Plato, it was the final cause in which Aristotle was most interested. And, like Plato, to him final causes are always teleological and explain what takes place in terms of fulfilling some end purpose. But, unlike Plato, he did not attribute purpose to participation in an external and independently existing set of Forms. Instead, Aristotle saw purpose as being immanent, as residing in the things themselves. There exists in every material substance a natural impulse or drive to achieve its proper form as perfectly as possible, without need of any external influence or further explanation. It is simply part of the inherent nature of the world, part of the governing logos, the same way that matter can combine with form to make an actual substance while still possessing the potentiality of forming other substances. Aristotle extended this same way of thinking to all physical phenomena and in particular to explanations of motion.

Thus a body moves not primarily because it is acted upon by other material bodies but to achieve some intended final result. Heavy bodies fall to earth because all things have an innate tendency to move as close as possible to the center of the universe, located at the center of the earth. The Greeks were already well aware that the earth was a sphere from observing, among other things, the shape of the earth's shadow on the moon during lunar eclipses and the shift in position of the stars as a person traveled north and south. The motion of a falling body is arrested at the earth's surface by the material substances of the earth, which have already aggregated uniformly in all directions about the center of the earth (the center of the universe) in the shape of a sphere. The heaviest and most dense substances are those closest to the earth's center, since they are able to displace the lighter less dense ones located nearer the surface. One finds granite and other dense rock far below the ground, while water is largely confined to the surface in rivers, lakes, and oceans or occupies voids in the earth as in the case of underground springs and pools. Air, being lighter than either earth or water, is displaced by both and is located in the region above the earth's surface. Fire, lighter still, rises through the air as does heated air and the steamy vapor from boiling water. The lightest and most tenuous substance of all, which the ancient Greeks referred to as the *aether* (sometimes spelled ether), is found in the heavens beyond the region occupied by air. Aristotle, like Empedocles, ascribed elemental importance to earth, air, fire, and water, adding the aether as a sort of rarified fire or luminous substance, an idea that was to persist over the years in our thinking about the nature of light and that was resurrected at the end of the nineteenth century in a desperate last-ditch effort to save then current theories about the propagation of light.

Unlike Heraclitus, Aristotle concluded that the natural tendency of all bodies is to be at rest. Rest then becomes their telos and natural state. He noted that an object, or a body, set in motion will eventually come to rest unless an external influence such as a push or a pull from another body acts upon it to keep it moving. Here we can distinguish between what Aristotle termed *natural motion* and motion caused by the action of an external agent. Natural motion in Aristotle's world takes on only two forms: motion in a straight line and circular motion. Natural motion in a straight line is either toward the center of the earth, as in the case of falling bodies, or away from the center of the earth, as in the case of matter moving upward. Straight line motion always ceases whenever a moving body has reached its proper place as close to the center of the earth as possible, as when a falling body comes to rest on the earth's surface or a lighter body is displaced upward by a heavier one. Straight line motion is not the permanent state of substances but is only a temporary or transitory condition required to achieve the end of having each body assume its proper place with respect to the center of the earth. Natural circular motion is neither toward nor away from the center of the earth but takes place in a circle about the earth's center. This is the motion of the heavenly bodies around the earth. It is the only kind of motion that can be eternal, since it produces no change in the distance of a body from the center of the universe; but for it to last eternally it must be caused by a continuously acting external influence, otherwise it too would die out.

The Greeks viewed circular motion as the only form of perfect motion, partly because it alone produced no change in the ordering of the universe and partly because it produced no change in itself. A circle rotating about its center is unchanged in any way. The circle was regarded as the most perfect of all plane figures, since it is the only one that is symmetric in every direction about its center. The sphere was the most perfect of all solid figures for the same reason. A universe that was ordered in the best of all possible ways would undoubtedly be one in which perfect motion would be circular about its center, and all eternal motions would of necessity be perfect. This notion of circular motion as somehow more fundamental than other kinds of motion was a Greek fixation and one that became firmly embedded in their thinking as a meta-principle. It is an easy one to understand, since it strongly appeals to our aesthetic sense of simplicity and beauty and perfection, but it is also one that shuts out other useful ways of thinking about the world. The view of the preeminence and perfection of circular motion was to persist largely unchallenged until the end of the sixteenth century, and even then it would die hard. It continues to shape the way we think of the world today.

There were other forms of motion besides straight line or circular movement, but all these others were caused by external influences, as when one body pushes against another, and were not considered part of the fundamental nature of the universe. Aristotle held that motion produced by an external agent ceased once the external cause was removed. This left him the embarrassment of having to explain how a javelin continued to move once it left the thrower's hand. He did this with much inconsistency by invoking another aspect of his physics, namely the impossibility of a void. On this point Aristotle sided firmly with the Eleatics. When a javelin moves forward, it displaces air from in front, which in turn displaces other air that flows in behind to occupy the region the javelin vacates as it moves forward. This air flowing in behind the javelin, he argued, is what continues to push it forward. He did not bother to explain how a javelin with a sharp point on both ends could be pushed as readily as one with a blunt or flat rear end, but that is only one of the many inconsistencies in Aristotle's treatment of motion. His explanation in fact is one that clearly admits the possibility of perpetual motion. A javelin propelled in this fashion could continue moving forward forever since there is nothing to stop the process Aristotle invokes.

This example linking sustained motion with the absence of a void illustrates the degree to which Aristotle's ideas about motion were tied up with the rest of his physics. To him, a void is impossible because space was not separate from matter. Space exists at all only by virtue of the material substance occupying it. Hence a void, or empty space, is not only impossible it is an oxymoron. The universe—that which exists—would be filled everywhere with matter as a prerequisite for its existence. If it weren't, it wouldn't exist at all. In this way Aristotle stands Heraclitus on his head. Motion does not create the material world but rather the other way around. Thus it is necessary to have the aether—or something material—to fill the farthest reaches of the universe. The matter that everywhere fills the universe creates a resistance to motion, which is why all motion ceases unless sustained by an external influence or, in the case of natural motion, by the inherent tendency of a body to seek its proper place with respect to the center of the universe. Of course this doesn't help at all to explain why the push received by the javelin from the air behind it is greater than the resistance it encounters from the air in front of it, as would be required for Aristotle's explanation of sustained motion to make any sense.

The earth must be at the center of the universe, in Aristotle's view, since if it were not, it would exhibit a natural motion toward that point like every other object. Yet the earth itself cannot be moving or a body thrown straight up would not come back down where it started and one dropped straight down would not land on the point directly below it. The earth cannot be spinning or rotating for the same reason, and because if it were the objects on its surface would be slung sideways like so many marbles placed on a carousel. The universe itself must be finite since it must have a center toward which natural motion is directed. If it were infinite in size and extended without limit in every direction, then any point within the universe would be equally distant from its farthest extent and would serve equally well as its center. Not only is it finite, it must be spherical in shape since everything moves equally toward its center from all directions and the distribution of matter as it tries to get as close as possible to the center would have taken on a spherical configuration. The question of what lies beyond its boundary does not arise. Since there is no matter beyond the boundary of the universe, there is likewise no space there, space being in Aristotle's scheme merely a property of matter. To the question of what is outside the spherical universe, Aristotle answered nothing, thereby forever banishing non-being from the material world.

Though the universe is finite in size, it has no beginning and no end in time. Time then is infinite in duration. We know time only from change, or motion, for which there must exist material substances. Thus there can be no time prior to when the universe existed and no time after which it will not exist. There can, says Aristotle with disarming simplicity, be no time before time was, or after it ceases. Hence time is everlasting.

Any careful observer of the sky would see the sun, the moon, and the "fixed stars" appear to move about the earth along circular paths that repeat in a fixed and periodic fashion from day to day and year to year. Thus it seems quite natural that to the ancient Greek mind the universe would be divided into concentric spherical regions or zones arranged like nestled spherical shells. The inner sphere, including the earth and the sub-lunary atmosphere and extending out to the sphere of the moon, is the region of change and decay, where substances can come into being out of matter and form or pass away to be transformed into other substances. All things in this region are made from the four elements earth, air, fire, and water and exhibit natural motion toward and away from the center as well as motion caused by external agents. To Aristotle, heat is the essential factor in generation, and he made the varying heat of the sun, as it moves along the path of the ecliptic on its sphere about the center of the earth, the ultimate efficient cause of all generation and decay in the inner sphere of the universe.

Beyond the inner region, from the sphere of the moon to the sphere of the fixed stars constituting the outer boundary of the universe, lie the incorruptible and eternal regions of the universe. This is the realm of perfection and permanence and here all motion is circular—eternal yet producing no change in the ordering of the universe. Here too along with the spheres of the moon and that of the sun are found the spheres of the five visible "wandering stars": Mercury, Venus, Mars, Jupiter, and Saturn. This outer region is filled everywhere with a fifth element, the luminiferous aether, a translucent and exceedingly fine and gossamer substance, unchanging and capable only of rotational or circular motion. The aether is necessary since there can be no voids anywhere in the universe. The rotation of each of the concentric spheres around the center of the universe (located at the center of the earth) is produced by the rotation of the sphere immediately beyond it. The aether is the mechanical means of transmitting the rotation of each sphere to the one inside it. The spheres must be in contact through some substance since there can be no voids and since, to Aristotle, motion could only be transmitted through mechanical contact.

The entire universe then is a kind of cosmic perpetual motion machine, with each of the concentric spherical shells kept in motion ultimately by the rotation of the outer sphere of the fixed stars, the *Primum Mobile*—prime mover—of the ancient astronomers. Since all motion will cease unless caused and sustained by something, Aristotle was left having to finally explain the perpetual motion of the outer sphere. This he did by postulating the Unmoved Mover, or First Cause, a variation on Anaxagoras's *Nous,* or Mind. The Unmoved Mover, which is the closest Aristotle comes to a deity in his philosophy, is, like time, eternal, with no possibility of change or motion. The Unmoved Mover is pure actuality and is literally immaterial. It cannot contain matter, since matter possesses the potential for change. Though producing motion, it is itself immobile, since there is nothing beyond it to serve as the cause of its movement. It is likewise immutable for the same reasons.

From this brief consideration, we can begin to appreciate how tightly coupled Aristotle's cosmology is with his physics and in particular with his ideas about motion. It is hard to dispute one without having to disagree at some point with the other until eventually the whole thing collapses. His cosmology is rigidly geocentric, with a stationary earth at the center and all of the heavenly bodies rotating around it. The same apparent motion of the heavens could be explained in a different fashion by the rotation of the earth about its axis. But then one would have to explain why bodies fall straight down instead of moving sideways as the earth turns beneath them, as required by Aristotle's dictum that natural motion is in a straight line toward the center of the earth. Disagreeing with Aristotle's views about motion is doubly difficult, since he only expresses what each of us has experienced for himself. A body does indeed appear to fall straight toward the point above which it is released, as if the earth is stationary below it. How could an object thrown straight up return to the same spot if the earth were turning beneath it? If the earth rotated once around its axis in twenty-four hours, as would be required to explain the apparent motion of the stars, a person at the equator would have to be traveling toward the east at more than a thousand miles an hour. Yet we experience no sensation of moving. Why are objects loose on the earth's surface not slung off, the way objects on a rotating carousel are? To remain at rest does seem to be the natural state of all bodies; witness that any object we set in motion soon comes to rest. To transfer motion from one body to another does seem to require physical contact between the two. Motion can only be sustained by continuing to push on a moving body. The world does seem to be filled with material substances everywhere since any medium through which we try to move a body resists its motion. Heavier bodies do fall faster then lighter ones, even if not exactly in the manner Aristotle claimed. The examples go on and on.

Aristotle's physics reflects a very natural and simplified view of the world, much like that possessed by a child or anyone looking seriously for the first time and reporting frankly what he sees. It relies almost entirely on direct sense perceptions of reality with few abstract concepts. Where an explanation is needed, it is entirely teleological in terms of some purpose the world is trying to achieve. The ball falls and smoke rises because both are trying to attain their proper place with respect to the center of the universe. There is no need to speak of gravity. Heavier objects merely displace lighter ones as everything falls toward the center of the earth. There is little question the earth is at the center of the universe, since otherwise it too would have fallen to the center had it not been there already. And of course the universe has existed for all time. How could all that there is have any beginning or end? To imagine it is beyond our comprehension. To express it is beyond the scope of language.

All of these are quite natural ways of explaining our direct experience of the world. They seem so evident to us that to dispute them would appear foolish. Teleological explanations are just part of that natural way of thinking about the world. We are creatures motivated by wants and desires, hopes and aspirations, to achieve which we become purposeful in our actions. We are accustomed to thinking about why we do things in terms of what we are trying to achieve. It is a manner of thinking that persists in our speech and in our colloquial explanations long after we have become more sophisticated in our knowledge and understanding of physical phenomena. In describing the motion of electric charges, the physicist is apt to say things like the electron "wants" to get as close to the proton as possible, or as far away from another electron as it "can"; or the electron "sees" the electric potential of the proton or "feels" the repulsion of another electron; or an excited atom emits a photon because it "needs" to reduce its total energy to the lowest allowable level. The physicist can afford to speak this way, because it describes qualitatively what happens and because he no longer relies on such explanations but has an entirely different quantitative description of events that allows him to make precise predictions about the outcome. Even so, the success of the mathematical description does not preclude this teleological and animistic way of thinking and talking about the world.

Aristotle's cosmology was not really as simple as we have made it sound, nor could it have really worked the way our brief explanation suggests. The difficulty arises from having to account for the observed irregular motions of the five visible planets. The Greeks named these bright objects *planets*, meaning wanderers, because they appear to move about with respect to the other stars. Normally, they move slowly from west to east against the background of fixed stars, but on occasion they can reverse direction, or retrogress, in their motion for a period of time before once again resuming their slow drift eastward. By the time of Aristotle, the motion of the planets had long been observed and was well known. Plato issued a challenge for someone to explain their behavior in terms of regular and ordered (meaning circular) motions. One of his pupils, Eudoxus, conceived a scheme involving a series of concentric spherical shells rotating at different speeds in opposite directions that was capable of replicating qualitatively the kind of retrograde motion observed for the planets, though it never quantitatively reproduced the observed motion. Aristotle employed the scheme of Eudoxus in his own cosmology, using a total of some fifty-five concentric spheres to account for the motion of the moon, sun, planets, and stars. The results were never more than approximately correct in predicting the observed motions. In addition, since some of the spherical shells had to move in the opposite direction from adjacent shells, it was hard to imagine how the motion of the outer sphere of the stars alone could produce all of the other motions called for.

Still, Aristotle's cosmology was the most complete and intricate ever attempted and at least on the surface sounded mechanically plausible. In addition it was intimately tied up with his physics and his ideas about matter, space, and motion. Even when the idea of actual material spheres made of aether was abandoned after Aristotle's death, his influence continued. The scheme that replaced the celestial spheres still adhered to his ordering of the sun, moon, and planets in the heavens and his ideas about the nature of motion. All heavenly motions had to be circular. Somehow the irregular movements of the planets had to be explained without resorting to any motion more complicated than movement along circular paths. The scheme that evolved became the Ptolemaic model of the universe that was to persist until well into the time of Galileo and his successors.

This scheme retained the concept of an outer sphere of the fixed stars, with the earth at its center. But the other heavenly bodies—the sun, the moon, and the five visible planets—were each envisioned to move along a smaller circular path called an *epicycle*, the center of which traveled along another circular path called a *deferent* with the earth at its center. All individual motions were still circular, but the circles were not all centered on the earth. The combined motion of the planet along its epicycle and the movement of the center of the epicycle around its deferent produced a resulting orbit of the planet around the earth that was itself no longer circular. This scheme accounted for *retrograde motion,* when the movement of a planet along its epicycle was in the direction opposite to the motion of the center of the epicycle along its deferent. It also accounted for the varying brightness of the planets by varying their distance from the earth and could be used to explain a number of other known features of solar, lunar, and planetary astronomy. There were in fact many different variations of the basic configuration depending on what effect one wished to reproduce. By adding more and more epicycles, and by using epicycle on epicycle, displacing the center of the deferent from the center of the earth (an *eccentric,* literally, out of the circle), placing the eccentric itself on a deferent, and other such variations, this scheme could be made to reproduce the observed motions more and more closely, but at the cost of ever-increasing complexity. The added complexity was the price that had to be paid for allowing only combinations of circular orbits for the motions of heavenly bodies.

The Greeks sought simplicity and perfection in the world, but often the simplifications they demanded cost them the simplicity they sought. Aristotle's insistence that all natural motion be either linear or circular is a perfect example of our capacity—even our tendency—to become so enamored with an idea that we categorically exclude all other possibilities and in so doing shut out other valuable ways of looking at the world. Sometimes such ideas are based on essential insights and turn out to be useful, as in the case of Newton's generalization of motion, Max Planck's quantum principle, and Einstein's principle of relativity. Then we exclaim over the genius that produced such insights and quickly forget about all the other possibilities that fell by the wayside because they turned out to be not so productive. Sometimes we become so blinded by one embodiment of a fruitful idea that we cannot recognize its correct application. In the motion of the planets, the simplicity and unity of organization that the Greeks sought was there all along and was to be eventually found in something over which the Greeks were the unparalleled masters: the conic sections of Euclidean geometry. These are curves formed by the intersection of a plane with a cone. But overly constrained, the application of the conic sections to motion was also overlooked, and its final discovery would have to wait for the developments of the seventeenth and eighteenth centuries.

Heraclitus made motion the centerpiece of creation. The Atomists suggested that everything was made by swarms of atoms moving chaotically through the void, colliding by chance and combining to form all material substances. By the time of Aristotle, little over a century later, the universe had been subdued and reduced to an orderly, well-regulated mechanism. From the rich diversity hinted at by Heraclitus we pass to the methodical, systematic world of Aristotle, in which motion has been constrained and confined and relegated to a relatively minor status. All natural motions are either in a straight line toward or away from the earth's center or are circular about the earth. The earth itself enjoys special status as the center of the universe and is at rest. Furthermore the natural state of all matter is to be at rest. Motion is no longer the defining characteristic of the world. Change does not depend on movement but is a result of the potentiality inherent in matter itself to exhibit various forms, allowing it to be transformed from one material substance to another. Most of Aristotle's physics is just restating the apparent in his more formal and logical sounding manner, without saying anything new. Wherever he needs explanations, Aristotle attributes what happens to tendencies exhibited by matter itself. Objects fall because material substances obey an inner impulse to move toward the center of the earth. This says nothing more really than objects fall because they do, which is a feature common to practically everything Aristotle has to say about physics. His physics is more in the nature of organizing what we observe and explaining it all as achieving what is best for the world. As for motion, there is very little to explain. Heavier objects fall toward the center of the earth displacing lighter ones that move away from it. Everything else circles eternally around the center of the universe. All other motions are the result of one object pushing on another in a universe completely filled with matter. This is how it must be since the earth is at rest.

After Aristotle, it would not be possible to raise the question of motion on the earth without at the same time bringing up the question of the earth's motion. However complex it might have to be, the Ptolemaic model could nevertheless explain the observed motions of the heavens as well as required. And it made use of nothing but the intrinsic simplicity and perfection of circular motions superimposed. Either to save appearances or to adhere to the principle that circular motion should be favored, it was unnecessary to look any further. The earth was securely at rest at the center of the universe, and earthly motion was then clearly as Aristotle had said.

And it was Aristotle's view—reasonable, conservative, reassuring, self-evident—not the paradoxical mutterings of Heraclitus or the wild, imaginative speculations of the Atomists that found favor and prevailed. Aristotle was the first truly systematic thinker, and his influence was both enormous and lasting. He brought order to a world beset by political turmoil and uncertainty and cultural decline, in which Athens and the Greek cities had lost confidence in the social institutions and accomplishments of the previous century. They wanted the reassuring certainty of answers, not more unresolved questions. And for almost two thousand years after Aristotle, motion was no longer considered a problem in need of a solution.

6

How versus Why

Aristotle asked why the ball falls and gave as his answer that every material substance tries to attain its proper place in the universe. It seems like a perfectly natural question. The aim of philosophy is to understand the ultimate reasons that govern the universe. That things should happen for a reason, and that there should be a purpose behind what takes place, seems like an equally natural assumption. Surely a world that has produced a creature that can ask such questions and have such thoughts cannot be without some guiding purpose. Our success and even our very survival depend on our ability to fathom reasons why things happen as they do. It is only natural that we would come to see events in terms of a purpose behind them. So Aristotle asked why things move and made his answer reside in the purpose of the motion.

Although it is perfectly natural to ask why something happens, it is not a simple matter to give a satisfactory answer. Questions that ask why cannot be answered in any absolute or unambiguous fashion. When we ask why, we are demanding to know reasons. If the reasons were apparent, then we wouldn't have to ask the question in the first place. Of course, it is easy enough to give some kind of reason. When asked why the light went out, we might reply that it was because the flow of electricity was interrupted or because the filament broke. Or, if asked why the plane crashed, we might reply because it ran out of fuel or the pilot made some error. To the question of why an object falls we might say, in the same way, because it was dropped. But these are only the immediate causes—in Aristotle's terminology, the material and efficient causes. Aristotle, however, is asking for the ultimate or final cause, which means always the purpose for which something happens, the desired end result to be achieved. And to give that kind of response to the question of why the ball falls one would literally have to understand the entire universe, the purpose behind everything, since the ball falling is not an isolated act but is related to everything else that takes place in the universe.

Aristotle is in essence asking one of the unanswerable questions of our existence. There is no way of responding to it that doesn't invoke more questions, each of which also asks why and demands as its answer a response that gives the ultimate reason or final cause. Each new question in turn leads to more questions, ad infinitum, and the inability to answer any one of them in an absolute fashion only ends up casting doubt on the answer to the original question. In response to Aristotle's answer, we could ask why the final resting place of the ball and not some other location should be its proper place, why material substances should exhibit this particular purpose and not some other, why the universe should exhibit any purpose at all, or even why the universe is here in the first place. The gamut of possible questions is limitless, and each new answer only opens up endless new questions.

Moreover, if we look critically at Aristotle's answers about why things move, we find that they really add little or nothing in the way of explanation to what we have already observed. They amount to saying, in effect, nothing more than objects move because they do. The explanation merely restates the observation that some things fall and other things rise or that the javelin continues to move until it comes to rest. The child's answer that the balloon rises because it wishes to get away, or that the stone falls because it wants to come down, is as illuminating as Aristotle's because it is exactly the same kind of answer. This apparent weakness turns out in retrospect to be one of Aristotle's strengths. His answer to why things move basically just says because it is the nature of things. But it is precisely the nature of things that we wish to know. And it is the nature of things, not the reasons why they are that way, that should be unambiguous. Given the same circumstances, the ball presumably falls in exactly the same manner each time we drop it. When we observe closely, we find that appears to be the case. This finding suggests that we shift the focus from trying to answer *why* an object falls to describing instead *how* it falls—that we try to understand the nature of the world by *describing* motion rather than attempting to *explain* it. We can go further and say that the only way to explain why things move is by describing how they move.

This is just the kind of shift in emphasis that began to take place in the seventeenth century as part of the renewed efforts to understand the nature of motion.

This renewed interest in motion grew out of the need to correct the calendar. For almost two thousand years after Aristotle, the understanding of the physical world changed very little and was based largely on Aristotle's ideas about matter and space and motion. The prevailing model of the universe was that of Ptolemaic astronomy with its close connections to Aristotle's physics. But the Ptolemaic model never quite worked. It couldn't adequately account for the observed motions of the sun and the planets and, in particular, the exact time each year of the vernal equinox. Revisions to the calendar were required periodically to keep the vernal equinox fixed at the same date each year. Since the calendar year is the same as the interval between one vernal equinox and the next, this meant that the exact length of the year was only approximately known. The time of the vernal equinox can be determined experimentally by observing the date when the sun rises and sets most nearly east and west.

Early solar calendars were based on a year of 360 days, which shifted the vernal equinox by more than five days each year. At some point, five extra days were added to give a year of 365 days, but after only forty years the vernal equinox had again shifted by ten days. In 45 BC, using the best advice available from the Ptolemaic astronomers, Julius Caesar revised the calendar based on a year of 365.25 days. The Julian calendar alternated three years of 365 days with a leap year of 366 days to achieve a four year average of 365.25 days. But the year is actually about eleven minutes and fourteen seconds shorter than 365.25 days, and, by the time of Copernicus in the mid-sixteenth century, the date of the vernal equinox had moved from March 21 to March 11. Thus in the Gregorian calendar, instituted by Pope Gregory XIII in 1582, leap year is suppressed three times in every four hundred years, which still leaves a correction of about one day every 3,300 years to be dealt with at sometime in the future. It was to understand the reason behind these discrepancies and how to go about correcting for them that the problem of celestial motion, at the instigation of the Church, was again taken up in earnest in the sixteenth century, notably by Nicholas Copernicus and also by Tycho Brahe, Johannes Kepler, and Galileo Galilei. But reconsidering the problem of celestial motion meant reconsidering the motion of the earth, and that meant reexamining the nature of all motion. And that in turn meant reexamining the entire edifice of Aristotelian thought, on which, by then, the Church had carefully constructed its philosophical understanding of the world.

Without ever articulating it plainly, or as such, and perhaps without even being consciously aware of it, the emphasis slowly turned away from trying to explain why things move as they do and became concerned instead with describing how they move. Thus Copernicus attempted to describe celestial motions using a heliocentric model with the sun fixed at the center of the stationary stars and the earth and the planets traveling around the sun in circular orbits. The apparent diurnal motion of the sun and the stars was accounted for by the rotation of the earth around its axis. This model gave a good qualitative account of the retrograde motion of the planets, which move around the sun in the same direction as the earth but which appear to move in the opposite direction against the background of fixed stars whenever the earth overtakes one of the outer planets in the solar system or is overtaken by one of the inner planets. But Copernicus's scheme was unable to account quantitatively for the observed motions of the sun and the planets using only circular orbits with the sun at the center. To make his model as good quantitatively as the various Ptolemaic schemes, Copernicus was forced to include epicycles and deferents and eccentrics, in short, all of the intricacies of the Ptolemaic system, until finally his scheme was fully as complicated and no more accurate than the Ptolemaic model it was meant to replace. Copernicus failed to solve the problem of the planets, but he did reopen the whole issue of the earth's motion about its axis and the possibility that the earth and the planets move around the sun. By challenging Aristotle's contention that the earth is at rest at the center of the universe, he challenged all of Aristotle's views about motion.

After Copernicus, literally on his deathbed, published his heliocentric scheme in *De revolutionibus orbium coelestium* in 1543, Tycho Brahe made careful measurements of the orbit of Mars that reduced the size of the experimental error in previous measurements by a factor of two. Using Brahe's data, Kepler, who himself had been working on the problem of planetary motion in the Copernican scheme for a number of years, finally concluded in a fit of profound dejection that the orbit of Mars could not be circular after all. Kepler lamented that for reasons surpassing all understanding, God had not employed the most perfect and simple of geometric figures for the orbits of the planets but had instead used a less perfect form, an *ellipse*, with the sun located at one focus. With that stroke the problem of the planets, but not the problem of motion in general, was finally resolved. So deeply entrenched was the idea of circular motion, to the exclusion of any other kind of path, that Kepler worked feverishly for a period of several years determined to make the orbit of Mars conform to some combination of circular paths. But such an assumption was not within the observational error of Tycho Brahe's data. It was beyond his belief and certainly beyond his comprehension, he said, that God would have resorted to using any path less perfect and hence less desirable than a circle. We will discover in time that Kepler had uncovered only a tiny piece of the puzzle. Celestial motions can in fact take the form of any of the conic sections from Euclidean geometry that were mentioned in the last chapter, but that ultimate realization would have to await the work of Isaac Newton. The underlying simplicity that Kepler sought was there all along, only it took a much different and far deeper and more encompassing form than he imagined.

Galileo's support of the heliocentric model eventually got him into trouble with the Church. The full nature of the dispute between Galileo and the Church was very complex, but it came to a head over the issue of whether the earth moved around the sun, or whether, as certain biblical passages suggested, the sun moved around the stationary earth. The Church freely granted permission for Galileo and other astronomers to make their calculations *as if* the earth traveled around the sun, so long as they did not try to argue that the heliocentric model represented the *actual* configuration of the heavens. The issue thus became one of modeling the world as if it were a certain way versus accepting the model as being an actual representation of reality. The Church, even in the face of mounting evidence to the contrary, tried to maintain a clean separation between these two uses of any mathematical model of the heavens, thereby holding on to a strict interpretation of scripture while granting the intellectual freedom to imagine and think about the world in other ways.

Galileo was not an astronomer in the mold of Copernicus and Brahe and Kepler, and he did not attack the problem of orbits directly by attempting to reconcile observations of planetary motions with the mathematical predictions of the various models. Instead he used a telescope fashioned from a simple arrangement of lenses to make observations of the moon, the sun, Venus, Saturn, and Jupiter and published his several findings in a little volume titled *Siderius Nuncius*, which it has become customary to translate as "Sidereal Messenger" rather than "Sidereal Message" as originally intended. Galileo used his observations of sunspots, the features of the moon, the phases of Venus, the rings of Saturn (which he could not fully resolve and so mistakenly interpreted), and the moons of Jupiter to argue in favor of a strict interpretation of the heliocentric model. He felt that his observations made the heliocentric model at least a plausible depiction of reality. But Galileo also realized that such indirect evidence, no matter how suggestive, would not be enough to counter the Aristotelian objections to the motion of the earth. He knew that he could not afford to be more adamant in his support of Copernicus's model of celestial motion without convincing arguments against Aristotle's views about motion on the earth. What was needed was a better understanding of motion in general. Galileo had been studying the motion of falling bodies and that of projectiles for a number of years. As a result of his virtual house arrest following his trial and recantation, he spent the remaining part of his life consolidating and compiling the results of his investigations, which were finally published in 1638 as *Dialogues Concerning Two New Sciences*. This work presented the first attempt at a systematic treatment and description of motion and marks the real beginning of mathematical physics.

In this work Galileo completely set aside any question of why bodies fall. It is a simple fact that they do. Instead he turned to understanding how bodies fall, in the expectation that answering that question would at the same time help understand eventually what makes them fall. Galileo started the study of motion down the path that physics has taken ever since: the only explanation of the world is to be found in the description of how things work. Questions asking why are deemed unanswerable. Physics begins with the expectation that questions asking how are not subject to the same limitation. We want the description of nature to be as exact and precise as we can make it, and that means it should be quantitative, formulated in terms of the numerical values of measured quantities. And it should describe the relationships that exist between the numerical quantities; in other words, the description should be mathematical. The physical quantities employed in the description will be those suggested to us by our immediate sense perceptions of the external world. Among them will be concepts like distance and space and time about which we already have an intuitive sense of their meaning. Others (and this will be increasingly true the more complete and encompassing the description becomes) may be far less intuitive—even counterintuitive or completely abstract and suggested to us only by the mathematical description itself and not by our direct experience of the physical world.

To achieve this exact and precise description, we will define things in terms of how they are measured. Distance, for example, we might choose to define as being what we measure with a meter stick and time as being what we measure with a clock, to take two specific instances and oversimplify them for the sake of illustration. As our understanding of the concepts deepens, we may find that we have to revise our definitions, but the rule will remain the same. The only meaning that any of the fundamental concepts will have resides in the specific operations by which we measure their numerical value. We term such definitions *operational definitions* and, beginning with Galileo, they are the foundation on which the description of the physical world is constructed.

This then is the *form* the physicist's world will take: a mathematical description of the relationships between fundamental concepts defined ultimately by how they are measured. What such a description has to do with reality of course resides in the physical meaning of the fundamental concepts and the mathematical relationships between them. These can be determined only by observing the physical world. The extent to which the description reflects the real world will always depend on our ability to discover what are the most meaningful concepts to use and our success in discovering the mathematical relationships linking them. These two things are of course connected, and success in discovering one is usually dependent on the other. The results of Galileo's experiments on the motion of falling bodies provides us our first concrete example of how the physicist's depiction of reality is constructed.

Galileo didn't actually perform his measurements on freely falling bodies. Such bodies quickly move too rapidly for the means of accurately measuring time and distance available to Galileo. Instead, he slowed the motion by letting the bodies move down an inclined plane. He realized that in both cases it was gravity that produced the motion and that bodies would move in the same fashion down an inclined plane as they would when falling to earth, only more slowly. For bodies moving down an inclined plane, the force of gravity is effectively reduced by the ratio of the height of the incline to its length. The longer the incline for its height, the more slowly the bodies will move. For our purposes, however, we will use the corresponding results for freely falling bodies.

When we set out to describe the motion of a falling body, we mean describing how its position changes with time. We use a very heavy, dense object so that we can ignore the effect of air resistance in slowing the fall of the body. The first concept with which we have to deal is the position or location of the body. For now we will assume that the motion is in only one direction—downward—and consider projectile motion later. To specify the position of the body, we have to measure how far it is from some reference point at any given time. For convenience, we will choose to measure distance from the point at which the body is initially dropped and time from the instant the body is released. As the body falls, we can measure how far it has traveled at the end of each second, for instance, and make a table of corresponding values of distance and time.

The numerical values in the table reflect what we experience as we watch the body fall. It starts off moving relatively slowly and then quickly speeds up. Each second the body moves progressively farther than it did in the previous second. It continues to speed up the farther it falls, but it is soon traveling so fast that our ability to discriminate by how much it speeds up during any interval becomes blurred. For that we have to rely on our measurements of the distance fallen during a given interval of time.

As we observe the motion, at least two additional concepts are suggested by what we find. The first is the idea of how fast the body falls, or its velocity. The second and related idea is the rate at which the velocity of the body is increasing, or its acceleration. We need to be clear about these two concepts and to distinguish carefully between them. The velocity of a body is the rate at which it changes its position, or how far it moves in a given amount of time. Acceleration is the rate at which the velocity of the body changes, how much the velocity increases or decreases in a given amount of time. From our table of measured values, we find that during the first second the body moves a total of 16 feet. We know that it isn't moving at a constant speed since it starts off at rest and accelerates, but over the entire one second interval the distance traveled is 16 feet, for an average velocity during the first second of 16 feet per second. During the next second we find that the body travels a distance of 48 feet, for an average velocity during that interval of 48 feet per second. The next second the body travels a distance of 80 feet, for an average velocity during that interval of 80 feet per second. We can use our measured values of elapsed time and distance traveled to compute the average velocity during each interval of time, just as we have done above for the first three seconds of travel.

Since each interval in the table is one second long, the average velocity in feet per second is numerically the same as the distance traveled in feet during that interval. From these values, we can see that the average velocity during each successive interval increases, reflecting the fact that the body is accelerating. But we also notice one additional thing. The increase in the average velocity from one interval to the next is constant throughout the motion and is equal to 32 feet per second. This is the amount by which the body speeds up each second. We express this by saying that the acceleration of the falling body is a constant 32 feet per second during each second, or 32 feet per second per second.

Thus we have used our direct measurements of how the object falls to arrive at an important characteristic of the motion: the object falls so that its acceleration is always constant. This finding becomes part of our description of how the body moves. It starts off from rest and accelerates at a constant 32 feet per second per second. During the first second its velocity increases by 32 feet per second, so that at the end of one second it is moving 32 feet per second. At the end of two seconds it will be traveling 64 feet per second, at the end of three seconds its velocity will be 96 feet per second, and so on.

We can describe this characteristic of the motion by a couple of mathematical expressions, namely

$$a = 32 \text{ feet per second per second}$$

and

$$v = at,$$

where t represents the elapsed time of fall, a is the label we use for acceleration, and v stands for the velocity in feet per second. A mathematical equation should always be thought of as a declarative sentence, in the manner of "John is tall." The equals sign represents the verb *to be*, or *is*. The subject is the portion of the equation to the left of the equals sign; the descriptor is the portion to the right. The first expression above, read as a complete sentence, states that the acceleration a is 32 feet per second per second (subject—verb—descriptor). The second expression states that the velocity v after t seconds is just the constant acceleration (the amount by which the velocity increases each second) multiplied by the number of seconds the object has been moving.

These two mathematical equations constitute our description of how the object we have observed falls when dropped. The first merely reminds us that the object falls with constant acceleration and gives us the particular value of the acceleration determined from our measurements. The second expression describes the velocity of the object at any time after it is released, not only at the particular times that we measured but at any other time as well. Our description says, for example, that the velocity one-half second after the object is released will be 16 feet per second, which is the same as the value we found for the average velocity during the first second of fall. Similarly, the velocity one and one-half seconds after the object is released will be 1.5 times 32, or 48 feet per second, which is the value we determined for the average velocity during the interval from one to two seconds. The average velocity of 80 feet per second we found for the third interval is the velocity at two and one-half seconds, and so on. We see then that the average velocity of the object during each interval of fall is the same as the actual velocity of the object at the midpoint of the interval. Or stating it another way, to find the average velocity during any interval we add the velocity at the beginning of the interval and the velocity at the end of the interval and divide by two.

If an object moves at a constant velocity of so many feet per second, then to find the distance it travels in a certain amount of time, say t seconds, we simply multiply its velocity by t. We can express this relationship in the form of another mathematical equation, $d = vt$, where d is the label we use for the distance traveled. In its entirety, this sentence (equation) reads: the distance d (in feet for example) traveled by an object moving at a constant velocity v (in feet per second) is the velocity of the object multiplied by the length of time t (in seconds) that it travels.

But a falling object does not move with constant velocity. Its velocity continually increases, so the distance traveled cannot be found by simply multiplying the velocity by the time of fall. To find how far it travels in a given amount of time, we can use the average velocity of the body, which is the total distance d the body falls divided by the total time of fall t, or $<v> = \frac{d}{t}$, where we use $<v>$ to indicate average velocity. Rewriting this equation, we have the distance given as $d = <v>t$. Now we found out that we can obtain the average velocity during the interval of time t by adding the initial velocity and the final velocity and dividing by two. The initial velocity (the velocity at the instant the object is dropped) is, of course, zero, and the final velocity is just $v = at$, so for the average velocity we get the value $<v> = \frac{0+at}{2} = \frac{at}{2}$. Notice that this is also the velocity at the midpoint of the interval, namely the velocity at the time $\frac{t}{2}$, just as it should be. Using this value for $<v>$, we can find the distance fallen as $d = <v>t$, so $d = \frac{at}{2} \times t$ or $\frac{1}{2}at^2$. Since in this particular instance, $a = 32$ feet per second per second, we could also write this expression as $d = 16t^2$.

We can now gather together our three mathematical expressions describing the motion of the falling object. They are

$$a = 32 \text{ feet per second per second,}$$

$$v = at,$$

and

$$d = \frac{1}{2}at^2.$$

Reiterating, the first expression states that the falling object accelerates at the constant rate of 32 feet per second during each second of fall. The second states that the velocity after t seconds is just the increase in velocity during each second multiplied by the number of seconds it has been falling and merely expresses the relationship between velocity and constant acceleration. The third expression gives the distance in feet the object has traveled t seconds after it is released.

Everything that we have observed, measured, and said about the motion is packed into these three simple equations. Once the symbols used in them are understood, our mathematical description of the motion carries with it not only quantitative precision but also, in its conciseness and brevity, great economy of expression. It captures the essential, measurable properties of the motion and expresses them as precisely and efficiently as possible. Each equation summarizes literally an infinite amount of tabular data, representing all of the possible numerical values of each of the symbols in it, and expresses in a single mathematical sentence the full range of our quantitative experience of the motion. One of the motivating ambitions of mathematical physics is to be able to describe more and more of our experience of the physical world by a single mathematical equation, to say in one succinct sentence how everything in the world moves and, in that description, to finally grasp the nature of reality. As limited as they may seem, these three simple equations describing the motion of falling bodies are an important first step in trying to realize that ambition. We will find in fact when we come to the work of Newton that these three equations can be reduced to a single all-encompassing equation of motion.

In claiming that these three equations constitute a mathematical description of the motion of falling bodies, we are making a number of assumptions that we should be clear about. In the first place, the description is based on the fundamental concepts of distance and time. By distance here we understand linear or straight line distance between some point on the falling body at any given time and the position of the same point before the body was dropped. Linear distance is measured in the usual manner by the use of a graduated standard scale, calibrated in feet or meters or some other convenient unit. By time, of course, we mean the intervals measured by a standard clock in units of seconds or fractions or multiples of seconds. Velocity and acceleration are derived concepts that depend only on distance and time and are likewise measured in units that are combinations of distance and time.

We also assume that if we repeat the measurements on which our mathematical description is based, using the same falling object, we will obtain the same results as before. In other words, we assume that the object falls the same way every time we drop it. How it falls never changes but is a fixed attribute of nature. This assumption about the constancy of nature is based on repeated measurements that consistently give the same results to within the expected variations in the measurements themselves. This is the same assumption that we make about causality, and although repeated trials may not constitute any kind of logical necessity for the repeatability of nature, our experience compels us to accept it as an operating principle.

A separate question is whether our mathematical description of how one object falls applies to other objects as well. Experiment shows that all sufficiently dense, heavy objects fall with the same constant acceleration over short enough distances, or what amounts to the same thing, over short enough durations of fall. The rate of acceleration varies somewhat with location on the surface of the earth but is constant at any given location. For lighter, less dense objects the rate of acceleration is less, and they do not fall as quickly as heavier objects. It is reasonable to suspect that the difference might be accounted for by the resistance of the air, and we find that in a vacuum (a void) all objects do fall at the same rate; a feather falls as fast as a lead ball. So we may conclude that, setting aside the effects of air resistance, our mathematical description pertains to the motion of all falling bodies. This was the conclusion reached by Galileo in studying the motion of bodies moving down inclined planes.

Galileo also studied the motion of projectiles and concluded that the motion of a body in the horizontal direction was unaffected by its motion in the vertical direction, and vice versa. Neglecting air resistance, a body moving in the horizontal direction does not change its velocity but continues to move at whatever velocity it was initially given. All of the acceleration of a projectile is in the vertical direction toward the center of the earth. The velocity it acquires and the distance it moves vertically in a given amount of time are unaffected by its horizontal motion. A body projected horizontally falls the same vertical distance in a given length of time as one dropped from rest. We can make use of this as Galileo did to determine the kind of path that a projectile follows in its motion.

A body projected horizontally with a velocity v_i (for initial velocity) will move a horizontal distance x during a given length of time t, where $x = v_i t$, since the distance traveled is just the velocity of the body multiplied by the length of time it travels. During the same interval of time the body will fall vertically a distance y, which from our previous description of falling bodies we know is given by $y = \frac{1}{2}at^2$. We can combine these two expressions to get a relationship between the vertical and horizontal distances traveled by the projectile, which will constitute an equation or mathematical description of its path. To combine the expressions, we must isolate t by dividing both sides of the first expression by v_i, $\frac{x}{v_i} = \frac{v_i t}{v_i}$, or t. Substituting this expression for t into the second expression gives us $y = \frac{1}{2}a(\frac{x}{v_i})^2$. We then rearrange its elements to isolate the horizontal and vertical distances, x and y, to arrive at the form $y = \frac{1}{2}(\frac{a}{v_i^2})x^2$. This last expression is the equation of the curved path called a parabola that resembles an upside-down U.

Notice that the vertical distance traveled by the projectile depends on the square of the horizontal distance, so that doubling the horizontal distance increases the vertical distance by a factor of four, tripling x multiplies y by a factor of nine, and so on. As a result the curve described by this equation changes faster in the vertical direction than it does in the horizontal direction. The result is that the path followed by a falling body projected horizontally is not straight but curves downward, just as we observe if we throw an object horizontally off a cliff and watch its path curve downward as it falls in the vertical direction. This equation gives us the exact shape of the path followed by the falling projectile. If an object is projected not horizontally but at an upward inclination, the path it follows is still a parabola as before but one inclined in the upward direction. The projectile rises for a while and then falls back to earth along its curved parabolic path. These were all conclusions that Galileo reached and confirmed by his observations of projectile motion.

Whereas Aristotle argued that natural motions like that of a falling body or those of celestial bodies had to be linear or circular, we see that Kepler discovered that the planets move in ellipses and Galileo that projectiles follow parabolic paths. Motion was slowly being unleashed from the constraints Aristotle imposed on it. Both of these findings were the result of observing how bodies actually move and describing the motion mathematically, rather than trying to answer why they move. Describing motion rather than explaining it—or rather, explaining it by describing it—was to be the direction taken by physics from the seventeenth century on. And in this sense Galileo is the grandfather of mathematical physics.

Our mathematical description thus far leaves a number of obvious questions unanswered. What is it, for example, that causes the acceleration of a falling body in the first place? If not matter exercising Aristotle's innate purpose, then what else? And why does the acceleration occur only in the vertical direction and not also in the horizontal direction? Why is the acceleration constant, and why don't heavier bodies fall faster than lighter ones the way we might expect from the weight we experience when lifting them? A description of how bodies move can never be anything other than that, namely a description, and can never answer every possible question about why things happen as they do. Nevertheless we can try to respond to these questions in the same manner—that is, to explain them by describing the motion; but to do so our description will have to become more detailed and complete and will need to incorporate some additional concepts. The most notable of these is the idea of a force or an external agent that can influence the acceleration of material bodies. Before we can deal with forces, however, we have to have a way of describing motion when the acceleration is not constant as it is in the case of falling bodies. In our description thus far, we have made use of the properties of constant acceleration in constructing the mathematical equations that relate how the velocity of the body and the distance fallen depend on the time of fall. These same equations do not apply to motion in which the acceleration itself is changing as the body moves. For the more general case, we will need the mathematical techniques of the calculus, which was invented by Isaac Newton for just this purpose.

7

Enter Newton

The world consists of change, said Heraclitus. His critics found such a world deeply disturbing. How could we hope to understand something that is constantly changing? By trying to understand change itself, the physicist answers. And for the physicist, the building block of change is motion—whether it is the motion of a body, or object, the ultimate parts of that body, a beam of light or energy from the sun, or the entire universe itself, wherever and in whatever form motion occurs. So the physicist sets out to describe exactly how things move, which means describing motion mathematically, in terms of its measurable properties. And for that it was necessary to know how to deal with quantities characterized not only by their value at any instant but by the rate at which they are changing. The man who first determined how to do this was Isaac Newton, the inventor, along with Gottfried Wilhelm Leibniz, of the calculus.

We experience the motion of a body as a change in its position or location, which we can specify by measuring its distance and its direction from some arbitrary and conveniently chosen point of reference. By making repeated measurements of the location of a body at fixed intervals of time, such as in the case of freely falling bodies, we can describe how the position of a moving body changes as a function of time, providing us with a quantitative description of the motion. Of course there are a number of practical problems involved in measuring the location of a moving body, such as how to determine where it is at some specific instant of time, since the object is moving even while the measurement is being made. But these are practical limitations associated with the way distance and time are measured and do not invalidate the method by which we go about describing the motion of the body. We could, for instance, imagine taking a series of photographs of an object and a clock simultaneously and using each photograph to determine the location of the object at the time shown on the clock. The location of the body could be measured by having a distance scale included in each photograph. The measurement of position and time would then be uncertain only by the length of time the camera shutter is open and that could be made progressively shorter and shorter by the use of more and more sensitive film. One can conceive of any number of other ways by which the motion of the body during the measurement could be suitably minimized.

We also experience motion in terms of how fast an object is moving, by its velocity or the rate at which it is changing its position. We speak of the number of feet per second or meters per second or miles per hour the object is moving. To determine the velocity of the body, we can measure how far it travels in a given interval of time and divide the distance traveled by the corresponding length of time. Thus if an automobile covers a distance of one hundred miles in two hours we can describe its velocity as 100 miles per two hours, or 50 miles per hour. If the automobile travels at a constant velocity during the entire two hours, then dividing the total distance traveled by the total length of time will accurately represent its velocity. But it might also have traveled at a number of different speeds during the two hour interval, in which case what we have calculated would represent only an *average velocity* during that period of time. This was exactly the situation that we encountered in the case of a freely falling body, which constantly increases its velocity as it falls. To more accurately determine the velocity at a specific time, we can measure the distance traveled during a shorter interval of time.

Suppose we wish to know the speed of the automobile one hour after the trip began. Then we might measure the distance traveled during the ten minute interval from 55 minutes to 65 minutes and divide that distance by the length of the interval to get the velocity of the automobile at one hour. Even during a ten minute interval the speed of the vehicle might change significantly, and the value we obtain for the velocity still might not reflect how fast it was actually moving at one hour. So we could shorten the interval even more and measure the distance traveled during the one minute interval from 59.5 minutes to 60.5 minutes, and divide that distance by the length of the shortened interval, to get a more accurate picture of how fast the automobile was moving 60 minutes into the journey. The speed of an automobile could still vary appreciably during even a one minute interval, so we might need to further shorten the time span to get an accurate value of its velocity at one hour. To guard against the possibility of any change in speed during the measurement, we would have to decrease the length of the time interval all the way to zero, which we cannot do since in that interval the vehicle would then travel no distance at all. As a practical matter, we can only make the measurement interval shorter and shorter, without decreasing it all the way to zero, in order to get a more accurate value for the velocity during the interval.

The procedure we have been describing suggests a general technique for determining the velocity of a moving body at a specific instant of time, which we term its *instantaneous velocity*. We measure the distance traveled during a given interval of time and examine the ratio of that distance to the length of the interval, as the length of the interval becomes shorter and shorter. If this ratio approaches a *limiting value*, that is, a value that it gets closer and closer to as the time interval approaches zero, we call this limiting value the instantaneous velocity of the body.

We need a convenient shorthand way of writing this procedure without having to say it in words every time. To see how we might do this, consider the example of a body that moves in only one direction so that its position can be specified by giving its distance from some fixed point of reference. Let y be the label we use to represent the value of this distance. In the case of a falling body, y would be the distance the body has fallen from the point of release. It is customary to employ the Greek letter delta (δ), sometimes shown in its capital form (Δ), to represent a change in some quantity. Thus δy represents a change in the value of y, or a change in the position of the moving body. We represent time by the symbol t as before, and let δt represent the change in time, or the interval of time corresponding to the change δy in the position of the moving body.

If an object moves a distance δy in a length of time δt, then its average velocity during the interval is just the ratio $\frac{\delta y}{\delta t}$. According to our definition of instantaneous velocity, we examine the value of this ratio as the length of the time interval is made shorter and shorter, that is, as δt approaches, but never quite gets to, zero.

Making use of our symbols, we can write this definition of instantaneous velocity in the form of a mathematical expression,

$$v = lim\left(\frac{\delta y}{\delta t}\right) \text{as } \delta t \text{ approachs } 0.$$

The symbols, $lim\left(\frac{\delta y}{\delta t}\right)$ as δt approaches 0, signify that we find the limiting value, or the value that the ratio $\frac{\delta y}{\delta t}$ approaches, as the time interval is made shorter and shorter. This mathematical sentence says succinctly in symbols everything that we have said at much greater length in words in describing the procedure by which we go about determining the instantaneous velocity; and all of the words are necessary to fully understand the correct meaning of the symbols. But once the meaning of the symbols is clear, then the mathematical expression becomes a much more efficient and compact means of expressing the procedure and the concept behind it.

This technique for determining the instantaneous value of a changing quantity is the fundamental idea behind what is known as the *differential calculus*, invented by Isaac Newton for just this purpose because, among other things, he needed it to formulate a mathematical description of motion. The calculus was to become the basic tool by which Heraclitus's world of change could finally be described quantitatively. Mathematicians call the quantity on the right side of our expression for instantaneous velocity a *derivative*, or rate of change; or more specifically in this case, the *derivative of distance (position) with respect to time (t)*. It is customary to further abbreviate it in some form, such as by the symbols $\frac{dy}{dt}$, or perhaps by $D_t y$. The latter is less confusing, since the former suggests erroneously that the derivative is a fraction, or the ratio of two quantities, though it is the form most often used. The derivative is not the ratio of two quantities; it is the value that ratio approaches as the denominator of the ratio becomes smaller and smaller. There are many subtleties associated with this whole notion of derivatives, including precisely what we mean by the concept of a limit and whether a limiting value for a given ratio actually exists or not. Exploring all of them in depth would take us too far afield and isn't really necessary for our purposes. Suffice it to say for now that the calculus has been the subject of intense scrutiny by mathematicians ever since its invention, particularly during the nineteenth century. The techniques and concepts have been made mathematically rigorous and today rest on a secure foundation.

For our purposes here, we will simply illustrate these ideas by applying them to the motion of falling bodies that we discussed in the previous chapter. There you will recall we described the motion of falling bodies by the three mathematical expressions;

$$a = 32 \text{ feet per second per second},$$

$$v = at,$$

and

$$d = \frac{1}{2}at^2,$$

where a is the constant acceleration, v the velocity attained, and d the distance traveled by a body falling from rest for a length of time t. In this chapter we will substitute y for d, because y is the conventional shorthand for vertical distance in mathematical equations. Suppose we start with the expression for the distance fallen and try to determine from it the instantaneous velocity of the body using the technique described above.

First, we need the distance fallen during some interval of time, say the interval from t_1 to t_2, where the subscripts 1 and 2 indicate two specific values of time. Let y_1 represent the distance fallen in time t_1 and y_2 the distance fallen in time t_2. Then, from our mathematical description of the motion, y_1 and t_1 are related by the equation $y_1 = \frac{1}{2}at_1^2$. Similarly, y_2 and t_2 are related by $y_2 = \frac{1}{2}at_2^2$. The distance traveled in the time interval between t_1 and t_2 is just the difference between these two values, or $\delta y = y_2 - y_1$ (read: the change in the position of the body is its position at t_2 minus its position at t_1); and the length of the time interval is just $\delta t = t_2 - t_1$. In the previous chapter, we used $<v>$ for average velocity. Earlier in this chapter we learned that it could also be written as $\frac{\delta y}{\delta t}$. So the two expressions are equivalent. The instantaneous velocity is the value that this ratio attains as δt becomes smaller and smaller or, in other words, as the two values of time t_1 and t_2 get closer and closer together.

We can use this ratio to find the instantaneous velocity. First let's examine the quantity $y_2 - y_1$. Using our previous equations, we can rewrite this quantity as

$$\delta y = y_2 - y_1 = \frac{1}{2}at_2^2 - \frac{1}{2}at_1^2 = \frac{1}{2}a(t_2^2 - t_1^2),$$

where we have taken the common factor $\frac{1}{2}$ out of both terms on the right. But $t_2^2 - t_1^2$ can be factored into $(t_2 + t_1)(t_2 - t_1)$. This can be easily verified by carrying out the multiplication. Thus the expression for δy becomes $\delta y = \frac{1}{2}a(t_2 + t_1)(t_2 - t_1)$, and the ratio $\frac{\delta y}{\delta t}$ becomes $\frac{\delta y}{\delta t} = \frac{1}{2}a(t_2 + t_1)$, where we have divided the common factor $(t_2 - t_1)$ out of both the numerator and denominator (recall that $\delta t = t_2 - t_1$). Now we have to find what the value of this ratio becomes as t_2 approaches t_1. By "approaches" we do not mean that time is traveling backward, rather that the two times are getting closer and closer to being the same. But clearly as t_2 approaches t_1, the quantity $\frac{1}{2}a(t_2 + t_1)$ approaches $\frac{1}{2}a(2t_1) = at_1$. Hence the instantaneous velocity at time t_1 is just $v = at_1$.

Thus the velocity of the falling body at any time t is just $v = at$, which agrees with the second equation in our original mathematical description. Starting with the equation for the distance traveled by a body falling from rest for a length of time t, we have used the techniques of the calculus to obtain the expression for the instantaneous velocity of the body at time t. These two expressions in our original description of the motion are therefore not independent; one of them implies the other. The link between them is supplied by the differential calculus. From the equation for distance, we can use the methods of the calculus to determine the equation for velocity. Let us continue and see if we can also determine the value of the acceleration in the same manner.

Above we saw that the velocity of the body at time t_1 is just $v_1 = at_1$. The velocity at a later time t_2 will be $v_2 = at_2$. The change in velocity during the interval from t_1 to t_2 is given by

$$\delta v = v_2 - v_1 = at_2 - at_1 = a(t_2 - t_1) = a\delta t.$$

Thus the average acceleration during the same interval is the ratio of the change in velocity to the length of the interval, or

$$\langle a \rangle = \frac{\delta v}{\delta t} = \frac{a\delta t}{\delta t} = a,$$

and hence we discover that the average acceleration is independent of the time of fall; or in other words, the acceleration of a falling body is constant. Of course the exact value of the constant acceleration has to be determined experimentally by actually measuring it, but the fact that it is constant is already built into each of the equations in our mathematical description of the motion.

Using the methods of the calculus, we can reduce our mathematical description of the motion of falling bodies to a single equation, the one giving the distance fallen as a function of time. Expressions for the velocity and the acceleration can be obtained from the relationship between the distance traveled and the time of fall. In the parlance of the calculus, we say that the velocity of a moving body derives from, or is the derivative of, its position with respect to time, or more properly, the *first* derivative of position with respect to time.

Likewise the instantaneous acceleration can also be expressed as a derivative. The average acceleration of a moving body is the change in its velocity divided by the length of time during which the change occurs, or $\langle a \rangle = \frac{\delta v}{\delta t}$, where δv and δt represent the corresponding changes in velocity and time. The instantaneous acceleration is the value that this ratio approaches as δt becomes smaller and smaller, or writing it symbolically as before,

$$a = lim\left(\frac{\delta v}{\delta t}\right) \text{ as } \delta t \text{ approaches 0.}$$

The instantaneous acceleration is the first derivative of velocity with respect to time. But velocity itself is the first derivative of distance with respect to time; so the instantaneous acceleration is the derivative of a derivative, or the *second* derivative of distance with respect to time. In the rest of the book, whenever we use the terms velocity and acceleration, we mean the instantaneous values of both quantities.

The techniques of the differential calculus enable us to start with a mathematical description of how the position of a moving body changes with time, and from that description calculate how the velocity and the acceleration of the body depend on time. This process can also be reversed. We can start with a mathematical description of how the acceleration depends on time and from it determine how the velocity of the body varies with time. From the resulting relationship between velocity and time, we can then determine how the position of the body varies with time. The techniques of this reverse process are no more difficult in principle than those of the differential calculus and constitute the subject of the *integral calculus* (described below), though they are often a good deal more difficult to implement in practice.

We can get some idea how the reverse process works by applying it to our mathematical description of motion for falling bodies. We begin this time with the expression for acceleration, which states that the acceleration of a falling body is constant throughout the motion. We divide the complete time of fall into increments of time, each of size δt. The incremental change δv in the velocity during each increment of time can be determined by taking the definition of acceleration as the change in velocity divided by the change in time, $a = \frac{\delta v}{\delta t}$. Multiplying both sides by δt, we get $a\delta t = \left(\frac{\delta v}{\delta t}\right)\delta t$, or $a\delta t = \delta v$. To find the total change in the velocity of the body during the complete time of fall, we add together all of the incremental changes in velocity. Adding together all of the individual increments of velocity, δv, gives us the total velocity v, which from the last equation is the same as the constant acceleration times the sum of all the increments of time δt, or just at; so we have

$$v = at,$$

the second of our three equations describing the motion of a falling body.

Next we want to use this equation for velocity to determine how the distance fallen depends on the time of fall. During each increment of time δt the body falls an incremental distance δy given by

$$\delta y = v\delta t = at\delta t,$$

where we have used the previous equation for velocity in writing the last part of this expression. Notice from the third expression that the increments of distance increase in size as time increases. Adding up all of the increments of distance on the left side of this equation gives us the total distance the body has fallen. But adding up the increments on the right side is a bit more complicated, because they are not constant but each one depends on the value of time. So we have to further break down the total time into smaller pieces. We can divide the total interval of time into n increments each of width $\delta t = \frac{t}{n}$, where n is an integer. By letting n become larger and larger, we can make the intervals of time smaller and smaller. We will use the integer index i to label these increments of time. The length of time that pertains to the ith increment is just $i\delta t = \frac{it}{n}$. Using this expression, we can write the incremental distance fallen during the ith increment as

$$\delta y = a\left(\frac{it}{n}\right)\left(\frac{t}{n}\right) = \left(\frac{at^2}{n^2}\right)i.$$

Now adding up all of the increments on the left side of this last expression will give us the total distance y the body has fallen. Adding up all of the increments on the right side means adding up all of the integers, i, from one to n. There is a simple little expression that gives us the sum of the first n integers, namely $\frac{1}{2}n(n+1)$. Try it when $n = 1, 2, 3, 4$, and so on. Substituting this result for i in the last equation and adding up all of the increments on both sides gives us

$$y = \frac{1}{2}\frac{at^2 n(n+1)}{n^2} = \frac{1}{2}at^2\left(1 + \frac{1}{n}\right).$$

Finally, if we let the value of n increase without limit so that the total interval of time is divided into smaller and smaller increments, then the term $\frac{1}{n}$ approaches zero and the last equation approaches the limiting value

$$y = \frac{1}{2}at^2,$$

which is the third of our equations describing the motion of falling bodies.

This example illustrates why the reverse procedure can be more difficult to implement in practice. The difficulty arises because the distance traveled during any interval of time is not constant but depends on the time of fall. As a result, we have to divide the total time span into smaller and smaller increments, during each of which the distance traveled changes so little with time that it can be treated as constant and the total distance traveled can be calculated by adding together all of the constant increments of distance.

In the differential calculus we begin with some quantity, like the position of a moving body, that is not constant but is changing in time and break it into smaller and smaller increments in order to pin down by how much it changes during each small interval of time and in that way determine its instantaneous rate of change, or as we say, its derivative with respect to time. Finding the derivative of some quantity is referred to as *differentiating the quantity*. In the integral calculus, we reverse the process and add together the small changes occurring during each increment of time in order to reconstruct how the whole quantity varies with time, a process known as integration, hence the name *integral calculus*. Both of these techniques, differentiation and integration, illustrate the essential insight of the calculus—that of dividing a changing quantity into incremental parts for which the change is small enough that we can deal with it quantitatively as unchanging during small enough increments. We break the changing world of Heraclitus down into smaller and smaller pieces until we have isolated what is changing and determined its behavior at each instant of time. From this knowledge, we then add the pieces back together to determine the behavior of the whole. There has been no more powerful—and no more practical—concept in the entire intellectual history of science and mathematics. That it works so well has had a deep significance for our understanding of the world. Why this is so will become more and more apparent as you get further into the book.

The utility of the calculus in describing physical phenomena like motion suggested that the material world was continuous and could be divided into ever-smaller portions, that it might in fact be infinitely divisible. This concept of a continuous, infinitely divisible reality behind appearances is the foundation on which the mathematical description of the physicist's world has been constructed. This view of the material world as continuously divisible persisted unchallenged from the time of Newton until the end of the nineteenth century. The twentieth century however has revealed that nature is not infinitely divisible but appears to be made up of discrete units of matter and energy, although we still do not yet know exactly what they are. What seems most likely, and what the physical theories of the twentieth century tell us, is that we will never know for certain, that this is just one more of the fundamental limits that we face in our efforts to describe the universe. There is even a suggestion that space itself, and time along with it, may likewise consist of fundamental, indivisible units; and this finding, surprisingly, comes from attempts to reconcile our understanding of the very tiniest with that of the largest and most remote regions of the universe. But even these discoveries of the discreteness of nature are the result of using the mathematics of continuity—the calculus—to describe the motion of material bodies. Armed with this new tool, Newton set out to describe how all bodies move.

8

The Laws of Motion

Isaac Newton invented the calculus, though he shares that claim with others, notably Leibniz; but to him alone goes the distinction of having used the calculus to discover how to describe all motion by a single mathematical equation. Newton's equation of motion was to become the model upon which all physics after the seventeenth century would be fashioned. If Galileo is the grandfather of mathematical physics, Newton is the father.

Newton was born in 1642, the year that Galileo died. He published his work on motion in 1686 in *Philosophiae Naturalis Principia Mathematica* (*The Mathematical Principles of Natural Philosophy*), at the urging of members of the Royal Society in England. Newton held up publication for close to two years while he tidied up a number of loose ends, but the *Principia*, as it is generally referred to, was the result of investigations that began when Newton was a student and continued during his tenure as Lucasian Professor of Mathematics at Cambridge University. Though rarely read anymore, the *Principia* remains one of the most important and influential documents in the history of Western civilization and the development of modern science. In it Newton expounded a philosophy of experimental or investigative science, brought together and systematized all that was then known about motion and added substantially to it, enunciated the three principles known as the laws of motion, disclosed experimental findings that allowed the concepts of force and mass to be placed on an empirical basis culminating in the equation of motion, proposed the mathematical law of universal gravitation, and applied his results to a wide range of problems including the motion of falling bodies, celestial motion, and the problem of the tides. One of his most impressive and convincing achievements was demonstrating that Galileo's description of the motion of falling bodies and Kepler's description of planetary motion were merely special cases of far more general principles that could be expressed by a single equation of motion. Newton's philosophy, his approach, his methods, and most of all his success in reducing motion to a mathematical description, determined the direction of science forever afterward and served as a powerful impetus for a strictly mechanistic view of the natural world. It is a view from which we are not likely to depart entirely, ever again.

The *Principia* is also important because it raised a number of crucial questions about the nature of space and time, even though Newton's concepts of space and time were incorrect and ultimately unsustainable. In the end they were replaced by the theory of relativity at the beginning of the twentieth century. But that's as it should be. Faced with an impossible task, that of trying to understand the universe, the only way one can make progress is by first being wrong. Every physical theory is born flawed. Usually the shortcomings are all too apparent right at the outset but are tolerated or set aside in deference to the revealing new insights that the theory provides, until in time the discrepancies become untenable and give rise to a new theory, equally flawed in some other way, in a process without end. Our current concepts of reality are just as inadequate, and they too will succumb eventually to the ravages of time and a changing perspective. There is not *the* way of understanding the world, only many different ways. That in essence is what the poets were trying to say before Plato banished them from his magnificent edifice of philosophy. Our interest in the *Principia* is as much for the unanswered questions it raises as for those few answers that it provides.

Our best entree to Newton's work is through the three laws of motion. Rather than stating them as Newton did (for one thing he wrote the *Principia* in Latin), let's express them in language more familiar to us.

> *First law:* Every body remains at rest or continues to move in a straight line at constant speed unless acted upon by an applied force.
>
> *Second law:* The rate of change of the momentum of a body is directly proportional to the force acting on it and takes place in the direction of the force.

Third law: For every action, there is always an opposite and equal reaction; or, the mutual actions of any two bodies are always equal and oppositely directed along the same straight line.

Before proceeding to discuss the three laws, we need to clarify a few points of terminology that we have been careless about up to now. The physicist makes a distinction (because nature does) between the terms velocity and speed, which we have so far ignored. That is because we have been talking mostly about motion in a single direction, as in the case of falling bodies, or about projectile motion for which we considered motion in the horizontal and vertical directions separately. Speed is the term we use to describe how fast an object moves, the rate at which it changes its position (in feet per second, meters per second, miles per hour, or some other suitable set of units). *Velocity* is the speed of a body in a particular direction and is an example of what we refer to as a *vector quantity*, one that has both a magnitude and a direction (e.g., 60 miles per hour due south). *Momentum*—the product of mass and velocity—is another example of a vector quantity, as is *force*—the rate of change of momentum with time. *Acceleration* is likewise a vector quantity, since it must specify the rate at which the speed of a body is changing and in what direction. Speed, however, is an example of a *scalar quantity*, one that specifies magnitude only. The speed of an object tells us the magnitude of its velocity (say in feet per second) but not the direction. For motion in only one direction, we often find the terms used interchangeably, though to do so may be confusing. Even motion in a straight line can be in either of two directions—forward or backward—and velocity must specify both how fast and in what direction a body is moving. Nature distinguishes between scalar and vector quantities; both the magnitude and direction of certain physical quantities have to be taken into account in describing their relationship to one another. Another point of confusion is the term *uniform motion*. To the physicist, the term designates motion in which some quantity remains constant. It is used sometimes imprecisely to mean motion at constant speed and at other times to mean motion at constant velocity (constant speed and direction), or even to mean motion at constant acceleration, as in uniformly accelerating motion.

Ever since Newton stated the laws of motion their actual content has been the subject of debate. Probably the one most singled out for criticism has been the first law, since it uses the term force without ever bothering to specify exactly what a force is or how it could be measured. Twentieth-century physicist Arthur Stanley Eddington criticized the first law as saying little more than bodies remain at rest or continue to move in a straight line at constant speed, except in so far as they don't. But in all fairness it says a great deal more than that. Each of us has an intuitive concept of a force as a "push" or a "pull," and the first law draws upon that intuitive notion. Since we are given no independent definition of what is meant by a force, we have to accept the first law as establishing the meaning of that concept. It defines a force as the externally applied agent that causes a body to change its state of motion—from being at rest or from moving in a straight line at constant speed. If a body undergoes any kind of acceleration, then we know that it is being acted upon in some fashion by an external force. Force then replaces Aristotle's innate purpose of matter to achieve a desired end result as the final cause of motion. Without forces there would be no motion; the first law tells us that even the body moving in a straight line at constant speed could not have been set into motion without a force to act upon it. Unlike Aristotle's innate purpose, forces are external agents. They are applied to a body rather than being an internal property of matter. Although the first law does not give us an operational definition of how to measure a force, it does at least specify what we mean by a force of zero. A body at rest or one moving in a straight line at constant speed defines the condition of zero force.

Establishing the concept of a force is only part of what the first law does. It also specifies that, in the absence of any forces, the natural state of matter would be to remain at rest. Given that bodies are set into motion by the occurrence of forces, then once the forces are removed the motion should continue in a straight line at constant speed. This is quite a remarkable assertion, since it clearly cannot be based on any actual observations or empirical evidence. It is instead a *thought experiment*, an attempt to understand the actual situation by imagining one that never occurs. Nowhere in the universe has anyone ever seen a body at rest or one moving uniformly in a straight line. Such situations simply do not exist in nature. The first law assumes that we could even know if a body was at rest or moving at constant speed in a straight line. But how would we know? At rest with respect to what? A body may appear to be at rest on the surface of the earth, but the earth spins about its axis and moves around the sun in an ellipse, the sun moves through the galaxy that we call the Milky Way, and our galaxy is receding from the other galaxies as the universe expands, and on and on. Likewise a body moving in a straight path on the surface of the earth at constant speed is actually moving with changing speed along a very complicated curved path produced by the curvature and rotation of the earth and its elliptical path around the sun, combined with the motion of the sun through the galaxy and the motion of the galaxy among the other galaxies. Finally, how could we determine whether a line is straight or not? With respect to what?

None of this was a problem for Newton. When he used the term "at rest," he meant it in an absolute sense. To Newton, space and time were absolute concepts. There was something in the universe that was absolutely at rest, with respect to which the position and hence the motion of all bodies could be determined conclusively. And that something was space. Space was the empty framework—the void—through which all material bodies moved. Space itself was at rest, and everything else moved through that fixed space. A straight line was one that was straight with respect to space itself.

It is one thing to hold such a view of space and quite another to offer convincing proof of it. How does one go about detecting this fixed space in order to measure the position of bodies in it and to measure the motion of bodies through it? Presumably one need only to find some one point that is fixed, some point that is absolutely at rest, with respect to which everything else is moving. We have already said that the earth won't do. Since the time of Kepler, the earth has been known to be spinning about its axis and moving around the sun along an ellipse. The sun won't do either since it is observed to be moving through the Milky Way. Nor will the center of the galaxy do since the Milky Way is receding from the other galaxies.

Newton did not know how to answer this question either, yet he was still convinced of the absolute nature of space. He was not aware of all of the celestial motions that we know about, many of which have been discovered or confirmed only in the twentieth century, in particular that of the stars and galaxies. So he did the only practical thing he could think of to identify what was at rest. He took the so-called fixed stars—those that were so far away from us they did not appear to be moving with respect to one another (hence the term "fixed")—and assumed that they were at least approximately at rest. The motion of all other bodies was to be determined by measuring changes in their position with respect to these fixed stars. Newton's solution of course does nothing to settle the question of whether the fixed stars are actually at rest or not. Even if they did not appear to be moving, there would be no way to tell if they were really at rest with respect to one another or only moving so slowly that their relative motion was undetectable. We now have ways of actually detecting their motion and know that they are not fixed as Newton assumed, leaving us no satisfactory way to defend the idea of absolute space and with it the idea of absolute motion. Later we will see how attempts to shore up these concepts led eventually to the theory of relativity. At the time, the only vindication of Newton's assumption about the absolute nature of space was that the laws of motion did seem to correctly describe the movements of material bodies with respect to the distant stars. Even now Newton's laws of motion remain a satisfactory description for speeds that are small in comparison to the speed of light.

As no one has ever observed the situation described in the first law, where do the concepts of a body at rest or one moving at constant velocity apply? According to the first law, they would apply to any region of space where there are no forces. And that is just the point. Since no one has ever observed a body absolutely at rest or one moving uniformly in a straight line, the first law tells us that forces exist everywhere in our world. We can have situations where locally the various forces acting on a body are balanced so that the net force is zero, as in the case of an object sitting "at rest" on a table. The force of gravity pulling the object toward the center of the earth is just balanced by the force of the table against the object pushing it away from the center of the earth. If these two forces were not balanced, the object would move in the direction of the greater force as required by the first law. Of course to an observer viewing the situation from somewhere in space the object sitting on the table is moving as the earth rotates about its axis and travels in its orbit about the sun, and so on, so there have to be other unbalanced forces acting on it, namely the gravitational attraction of bodies other than the earth, such as the sun. Remember that forces are vector quantities. When we combine them to find the total, or resultant, force, we have to take into account both the direction and the magnitude of the forces. To combine two forces acting in the same direction, we add their magnitudes. To combine two forces acting in opposite directions, we subtract the magnitude of the smaller from the greater to get the resultant force acting in the direction of the larger force.

We can also create a situation that locally gives the appearance of being force free. As Galileo correctly concluded, all freely falling bodies accelerate at the same rate in a uniform, or constant, gravitational force. So in any region where gravity is the only force acting and everything is in free fall, all objects fall in unison and appear to be at rest with respect to one another. If everything is in free fall, nothing will appear to be moving to an observer falling along with everything else. Now suppose this same observer gives one of the freely falling objects an initial velocity in some direction, say, by throwing it. We can treat the two motions of the object—that of free fall and that imparted to the object by throwing it—independently (remember Galileo's treatment of projectile motion, dividing it into separate horizontal and vertical motions). Once the object has left the observer's hand, it will no longer have a force acting on it in that direction and will, according to the first law, continue to move in a straight line at constant speed. The observer will be unaware of any motion in the direction of free fall, since everything is falling together at the same rate, and will only observe the movement of the object in a straight line at constant speed. An outside observer—one not in free fall but watching the motion from some vantage point in space—will see both motions, that due to gravity and that produced by throwing the freely falling object. To that observer, the motion of the thrown object will not be in a straight line at constant speed but will follow a curved path at increasing speed just like Galileo's projectiles.

So any local region of space where gravity is the only force acting and where everything (including observers) is in free fall will appear to all observers in that region to be force free, and the first law will appear to correctly describe the natural state of all motion. We call such regions *inertial frames of reference*, and it is in such reference frames that Newton's laws of motion describe the observed motion of material bodies. The inside of an elevator in free fall down its shaft would constitute an inertial frame of reference (if we could ignore the effects of any other forces, such as that of friction). The occupants inside would feel weightless, or force free, because the floor of the elevator would no longer push up against the force of gravity (their weight) pulling them downward. Anything they happened to take out of their pockets and place in the space inside the elevator would remain at rest where they placed it, since it too is falling at the same rate as everything else inside the elevator, or would move at constant speed in a straight line in whatever direction it was launched.

We have all observed an inertial reference frame in the pictures sent back to earth from the various orbiting space stations. There everything is in free fall around the earth. The motion of an earth satellite in a circular orbit is actually a combination of two motions. One is motion at constant speed in a straight line tangential to the direction of the orbit, as required by the first law since there are no forces acting on the satellite in that direction (again ignoring the effects of friction from the few molecules of gas at that height and ignoring the smaller gravitational force on the satellite from other bodies such as the sun and the moon). The other motion is free fall toward the center of the earth due to the gravitational attraction of the earth. The combination of these two motions—perpendicular to one another—is what produces the circular orbit of the satellite around the earth. We can describe the combined motion as that of free fall *around* the earth. Since everything traveling along with the space station is in the same free fall condition, it represents an inertial reference frame just like the inside of the freely falling elevator. Objects in orbit are weightless since all are falling together and nothing, including the space station, pushes against anything else. We have all watched fascinated as an astronaut places some object motionless in space and it just sits there unsupported by anything or moves slowly in a straight line in whatever direction it might happen to be launched, all in perfect obedience to the first law of motion. The situation described by the first law, which Newton could only imagine as a thought experiment, we can now observe routinely in the "weightlessness" of such freely falling reference frames.

Newton also considered time to be an absolute concept. By that he meant that there was a universal measure called time that pertained to the entire universe and did not differ from one location to the next. The present moment in one location was the same present moment in every other location. Another way of saying it would be that all parts of the universe would be the same age, have had the same amount of time in the past, and would have the same future. In his own words, "Absolute, true, and mathematical time, of itself and from its own nature, flows equably without relation to anything external." He distinguished absolute time from "relative, apparent and common time" by saying that the latter "is some sensible and external . . . measure of duration by the means of motion, which is commonly used instead of true time, such as an hour, a day, a month, a year." In other words, there is a time that applies to the universe as a whole. We can measure portions of it locally by means of motion, such as the diurnal and annual motions of the earth and sun and stars or the periodic motions of a mechanism in a clock. Local time is what we measure with a clock, but the absolute length of time, or duration, measured by a given clock would be the same everywhere in the universe. The clock would measure the same elapsed time here as it would anywhere else because it is measuring a portion of the universal or absolute time. This concept of absolute time is surely one that agrees with our intuitive understanding of what we mean by time.

Similarly Newton distinguished between local, or relative, space and absolute space. Local space was what we measure with a graduated scale, such as a meter stick or a yardstick, and might correspond to the distance of an object from some reference point or the volume occupied by the object. Local space could change or move, as objects changed shape or moved, but absolute space was immovable and fixed. In Newton's description, the local space measured by a distance scale would be the same in all parts of the universe, since it measured a segment or portion of the absolute fixed space. As we have seen in the case of the first law, the concept of an absolute space, fixed and at rest, with everything else either at rest in it or moving through it, is essential to the meaning of Newton's laws of motion. The concept of an absolute or universal time is equally essential, since motion involves the measurement of elapsed time, or duration. If measured time, like measured space, were not the same everywhere, then the laws of motion could not be universal.

The second law further elaborates the concept of a force. It equates the force acting on a body to the rate of change in its momentum. Furthermore, it tells us that the change in momentum is in the same direction as the force. By the momentum of a body, Newton meant the product of its mass and its velocity. We could take the second law as a quantitative definition of force, if only we had an independent definition of what we mean by the mass of a body, since we already have an operational definition of velocity. The rate of change of the momentum would give us the magnitude of the force and the direction in which the momentum changes would give us the direction of the force. Unfortunately, Newton provides us no definition of mass other than what the second law states. But we can't take the second law as a definition of mass, since we have no separate quantitative definition of force, other than the concept of a zero force given by the first law. So it looks like we are stuck. Somehow we must come up with an independent way of defining either force or mass in order to use the second law as the defining relationship for the other quantity.

To resolve this impasse, we have to start with the empirical observations on which the second law is based. Suppose we take two bodies—label them 1 and 2 for convenience—and let them interact, by setting them in motion for example and letting them collide or by connecting them together with a spring then stretching or compressing the spring and releasing them. During the interaction between the two bodies, they exert a force on one another. They push or pull against each other, either directly by colliding or through the intermediary of the spring, and it is this force of interaction that causes each to change its momentum. Since the quantity of matter in each body is fixed (these are solid bodies for instance), then the change in momentum shows up as a change in the velocity of the body. As a result of the interaction between them, each body changes its velocity, or is accelerated. For simplicity let the acceleration occur along a single direction.

Now suppose we carefully measure the acceleration of each body produced by the interaction between them. What we find is that in every case the ratio of the acceleration is constant, that is,

$$\frac{a_{12}}{a_{21}} = m_{21},$$

where a_{12} represents the acceleration of body 1 due to its interaction with body 2, a_{21} is the acceleration of body 2 due to its interaction with body 1, and m_{21} is a constant. In addition we find that a_{12} and a_{21} are always oppositely directed—the two bodies are accelerated in opposite directions. We give the constant m_{21} a name and call it the mass of body 2 with respect to body 1. Similarly, if we let body 1 interact with a third body—call it body 3—we find that

$$\frac{a_{13}}{a_{31}} = m_{31},$$

where again a_{13} is the acceleration of 1 due to its interaction with 3, and a_{31} is the acceleration of 3 due to its interaction with 1. In like manner we term m_{31} the mass of body 3 with respect to body 1.

Up to this point we have let bodies 2 and 3 each interact with body 1. Next we let them interact with each other. If we let body 2 interact in the same fashion with body 3, we find as before that

$$\frac{a_{23}}{a_{32}} = m_{32},$$

but in addition, as a separate empirical observation, we also find from our measurements that

$$m_{32} = \frac{m_{31}}{m_{21}}.$$

The mass of 3 with respect to 2 is just the ratio of the masses of the two bodies, each determined with respect to body 1. Using this last relationship, we can rewrite the one before it as

$$\frac{a_{23}}{a_{32}} = \frac{m_{31}}{m_{21}},$$

or by cross-multiplying, we get

$$m_{21}a_{23} = m_{31}a_{32}.$$

This final expression summarizes all of the empirical relationships that we have used and allows a particularly meaningful interpretation. The only value of mass that shows up on each side of the equation is the mass of each body with respect to body 1. The mass of either body with respect to the other—the values of m_{23} or m_{32}—do not enter into the final expression. This means that, if we measure the mass of a body with respect to some conveniently chosen standard body, like 1, we obtain a constant, in this case m_{21} or m_{31}, whose value is then independent of any other body with which the one in question interacts. This constant property of the body (with respect to the standard body) that we call its mass determines how the motion of the body changes, or how it accelerates, when it interacts with any other body.

We can rewrite the final expression above once more, this time in the form

$$m_2 a_2 = m_3 a_3,$$

where we have dropped the subscript 1 from the mass, since it is understood that both values are to be determined with respect to the same standard body. We can also drop the second subscript on the acceleration, since it is understood that the acceleration of each body is due to its interaction with the other. It is also understood that the acceleration of each body is in the direction opposite the acceleration of the other body.

This procedure is the origin of the standard kilogram. We choose a precisely specified volume of some standard material and arbitrarily designate it as having a mass of one kilogram. Then in principle we can let any other body interact with the standard kilogram and measure the acceleration of each. The mass of the body will be given by the empirical relation that we began with, $m = \frac{a_s}{a}$, where a is the acceleration of the body and a_s is the acceleration of the standard kilogram. This expression is just a special case of the final expression we obtained, $ma = m_s a_s$, in which the mass of the standard body has arbitrarily been assigned a value of 1. Thus we have a way, in principle at least, of measuring the mass of any body by comparing its acceleration with the acceleration of an arbitrarily chosen standard when the two interact in such a way as to change the motion of each. There is a much simpler way in practice of measuring the mass of a body that we will come to later.

Bear in mind that this procedure is not completely arbitrary but depends on the empirically determined relationships that we have used. In other words, it isn't merely something that we have concocted or made up but is based on measurements of how material bodies actually behave when they interact and undergo changes in their motion. The only arbitrary feature is that we are free to choose which is to be the standard body and to arbitrarily designate its mass as having a value of one kilogram. After that, the mass of every other body is fixed and is determined by its interaction with the standard mass.

This prescription supplies the operational definition of mass that was missing from our statement of the second law previously and incorporates the empirical observations on which the formulation of the second law is based. Having accomplished that, we can now use the second law as our quantitative definition of force. The value of the force acting on a body is given by the rate of change of its momentum. The momentum of a body is the product of its mass and its velocity, or in symbols, mv. If the mass of a body is fixed, meaning that the body is not losing or gaining mass by accumulating or losing material (such as an automobile or a rocket ship that is burning fuel and decreasing its weight), then the change in its momentum is due entirely to changes in its velocity. In that case, the rate of change in its momentum is just the mass times the rate of change in its velocity, or the mass times the acceleration, ma. By the second law, the product of mass and acceleration becomes our quantitative definition of force, or symbolically,

$$F = ma,$$

which is Newton's celebrated equation of motion.

Part of the genius of Newton is that he stated the second law in its more general form, recognizing that the force acting on a body is given by the rate of change in its momentum, including those cases in which the total mass of a body might be changing as well as its velocity. As a rather obvious example, imagine a conveyor belt on which material is being continuously dropped. The increase in the total mass of the conveyor belt produces a force acting on it that will cause it to slow down. Keeping it moving at the same speed requires the addition of an equal force in the direction of motion, which accelerates the material falling on the belt to the speed at which the belt is moving. For our purposes, we will limit ourselves to those cases where the total mass of a body is fixed and use the second law in the form given above, namely the force acting on a body is equal to the product of its mass times its acceleration, $F = ma$.

From this expression, we notice at once that the first law is contained in the second law as a special case. For if we let the force acting on a body be zero in the previous expression, we obtain $0 = ma$, and the acceleration of the body must likewise be zero. But if its acceleration is zero, then the body cannot change its velocity. It must either remain at rest or continue to move in a straight line at constant speed, just as the first law states.

Conversely, if a body does not accelerate, then the force acting on it must be zero. The second law is required therefore to completely understand the meaning of the first law, which is the main reason why Eddington and others criticized it as saying essentially nothing on its own. It is not actually a separate law but is only a special case of the second law.

While we are at this point we might as well go on to mention the third law. In our previous discussion we observed that when two bodies (say 2 and 3) interact, they change their motion according to the relation

$$m_2a_2 = m_3a_3,$$

where m_2 and m_3 represent the mass of each body and a_2 and a_3 represent the acceleration of each. We observed also that a_2 and a_3 are always in opposite directions. The product of the mass and acceleration of a body is just the force acting on it, so we can rewrite this expression as

$$F_{23} = F_{32},$$

or the force on body 2 due to its interaction with body 3 is equal (and opposite in direction) to the force on 3 due to its interaction with 2. The double subscript has been reinstituted to make the meaning of the symbols clearer. This equation expresses the basic content of the third law: that for every action there is always an opposite and equal reaction. Hence the third law is also contained in the empirical observations that we made use of earlier.

The term *action* in the third law refers to the force acting on a body due to its interaction with another body, either by colliding with it or in some other way being pushed or pulled by it. The *reaction* signifies an equal and opposite force acting on the other body. Perhaps we should emphasize that all forces acting on material bodies arise from interactions with other material bodies, that is, they are produced in some way by the presence, and motion, of other material bodies. The third law tells us that the forces between interacting bodies always occur in equal and opposite pairs. This observation is one of the unique findings that Newton added to our understanding of motion and is often cited along with the law of universal gravitation as his most significant contribution. As we have seen in our development of the laws of motion from the empirical observations that they represent, it is not quite so easy to separate the third law completely from the second law and to single it out as an independent contribution. Nevertheless it represents a separate fundamental and important insight into the nature of motion.

Newton had many simple straightforward arguments supporting the third law. A book at rest on a table pushes down against the table with a force equal to its weight. If the table did not push upward against the book with an equal force, then the combination of book and table would accelerate in the direction of the unbalanced force, as required by the second law. The same argument can be applied to the table and the floor of the house and to the house and the earth. If any of these forces were not opposed by equal forces in the opposite direction, the entire earth would accelerate (and thus constantly gain speed) in the direction of the unbalanced force, something that we simply do not observe happening or find any concrete evidence to support.

By the third law, internal forces—those occurring between the individual parts of a material body—must always occur in equal and opposite pairs. Consequently the total *internal* force acting on a body must always add to zero; the opposing pairs of equal forces cancel out. Suppose for a moment that were not the case. We can always think of an extended material body as being divided into smaller pieces of matter. If the internal forces between these individual pieces did not cancel out, there would be a net force acting on the entire body, and by the second law the body would spontaneously accelerate and gain speed forever in the direction of the unbalanced force. Again that is something we do not ever observe happening.

Gravity is a universal force of attraction between all material bodies. Imagine two objects in contact. Each of them exerts a gravitational force of attraction on the other. If these forces were not equal and oppositely directed so that they canceled each other out, then the two objects in contact, by the second law, would accelerate and move off in the direction of the larger force. Of course if the two objects are not in contact, such as an object suspended above the earth and then released, the gravitational attraction between them will cause them to accelerate toward each other, as in the case of an object falling toward the earth. Since the total force between the earth and the falling object is zero (equal and opposite forces), then by the second law the change in the total momentum of the two must be zero also. Whatever momentum the falling object acquires must be balanced by an equal and opposite momentum acquired by the earth. We do not, however, notice the movement of the earth due to its much greater mass. If we equate the momentum of the two at impact, $MV = mv$, where M and V represent the mass and velocity of the earth, and m and v represent the same quantities for the falling object. Then, dividing both sides of the equation by M, we see that at impact the earth would have acquired a velocity given by $V = \frac{mv}{M}$. The denominator of this ratio is so much greater than the numerator, due to the large value of the earth's mass, that the earth's velocity, though real, is imperceptible.

In each of the examples cited above the third law is required by the second law to explain why we do not observe spontaneous acceleration produced by internal forces or forces of contact. Newton's insight was to realize that the third law is universal and applies to the mutual forces of interaction between any two bodies. The third law is often a source of confusion as indicated by the following question: if the cart pulls back on the horse with the same force that the horse exerts on the cart, as required by the third law, how can the horse ever move the cart forward? The confusion results from failing to take into account all of the forces involved. The horse and the cart do indeed exert equal and opposite forces on each other. The horse and the cart both move forward, not as a result of the forces between each other, but because of the force that the horse exerts by pushing backward against the earth. If there were no friction between the horse and the earth, as would be the case on a perfectly smooth frictionless surface, then the horse could not exert a forward force by pushing against the earth and neither the horse nor the cart would move.

In thinking about the laws of motion, we made use of certain empirical observations to give us an operational way of determining the mass of any material body by comparing changes in its motion to those of a standard body when the two interact. In this scheme, mass is defined in terms of the set of operations by which its value is measured. We then used this operational definition of mass to define what we mean by a force. The force acting on a body is the product of its mass and its acceleration. We could have gone about it the other way around. We could have first defined what we mean by a force and used that definition to define mass.

Suppose we adopted as our operational definition of force the amount of stretch or compression produced in a spring of specific dimensions made out of some standard material when the spring is used to pull or push another object. An elongation or compression of a specified amount would be used to define what we mean by one unit of force. An elongation or compression twice as great would correspond to two units of force, and so on. We can then use this spring to pull or push on an object and measure the force applied to the object by measuring the extension or compression of the spring. The applied force will cause the body to accelerate if it is free to move. Now if we choose the material and dimensions of our spring properly, we find that for suitably small amounts of stretch or compression, the ratio of the applied force to the acceleration it produces is constant for any given body. Symbolically, $\frac{F}{a} = constant$, where F stands for the value of the force (measured by the change in length of the spring) and a is the value of the acceleration. If we increase the force, the acceleration is increased in the same proportion; the ratio of the two remains constant. This constant ratio appears to be a property of the body being accelerated. If we repeat the measurements with a different body, even one made of the same material, then in general we obtain a different value for the ratio, but once again the ratio is constant as we change the amount of force applied to the body. We give this constant ratio of applied force to the acceleration it produces a name and call it the mass of the body. If we double the volume of a given material, we find that the value of its mass, determined in the same fashion, also doubles, supporting the idea that the mass of an object measures the quantity of material, or matter of a given kind, contained in the body. Letting m represent the mass of the body, we can write our last expression as $\frac{F}{a} = m$, or $F = ma$, which is just Newton's equation of motion expressing the second law that we arrived at earlier. This time however we started with an operational definition of force, and the definition of mass arises naturally from the empirical observations that represent the second law. This approach gives us a more intuitive grasp of both force and mass. Force is the magnitude of the push or pull applied to a body; mass represents the inertia of the body, its tendency to resist changes in its motion. The larger the mass, the greater the applied force necessary to produce a given acceleration.

Mass and weight, though related, are entirely different concepts. Mass is an intrinsic property of matter, whereas weight is not. Weight is a force. On the earth the weight of a body is the gravitational force of attraction between it and the earth. Even in free fall, where a body appears weightless, it still has the same constant value of mass and still obeys the second law of motion. Whenever we arrest the fall of a body by holding onto it in some fashion, then we feel the gravitational force between it and the earth that we call its weight. If we suspend it from our standard spring, we can measure its weight by measuring the amount by which the force acting on it stretches the spring. If we release the body it will accelerate, and Galileo determined that it accelerates at a constant rate. The force causing it to accelerate is just its weight. The weight and mass of the body are thus related by the second law,

$$W = mg,$$

where W is the force acting on the falling body (its weight), m is its mass, and g is the value of the constant acceleration due to gravity. An alternative, and much easier, way of determining the mass of a body is to measure its weight and divide by the constant acceleration of gravity, which we can also measure. Since g is the same for all bodies, independent of mass, it has to be measured only once. A balance scale does essentially the same thing. It determines the number of standard masses (in kilograms or in fractions and multiples of kilograms) required to just equal, or balance, the weight of a body, giving its equivalent mass directly.

Newton's laws of motion are a synthesis of enormous insight and invention. They provide a succinct framework of principles built around the minimum number of physical concepts required to understand and describe the empirical observations of motion. They give quantitative meaning to the concepts of mass and force, and these, combined with the techniques of the calculus to define what we mean by instantaneous acceleration and velocity, provide a quantitative description of motion summarized by the second law. Newton's equation of motion becomes the universal description of how bodies move when acted upon by forces. In the Newtonian scheme forces, and not some Aristotelian purpose, are the final cause of all motion. Forces are the physicist's only answer to *why* things move, and it is an answer wholly grounded in our mathematical description of *how* they move.

If we know the forces acting on a body and its mass, the equation of motion specifies how the body accelerates with time. As we learned in our discussion of falling bodies, knowing how the acceleration changes with time, using the techniques of the integral calculus, we can determine how the velocity of the body varies with time; and from that we can determine how its position changes with time. Combining the equation of motion with the techniques of the integral calculus—both discovered by Newton—allows us to construct a complete mathematical description of how the body moves—its acceleration, velocity, and position as a function of time. The process works equally well in the opposite direction. Using the techniques of the differential calculus and the equation of motion, we can discover the kinds of forces at work in our world. If we know the position of the body as a function of time, we can take the derivative of position to determine the velocity of the body; then we can take the derivative of the velocity to determine its acceleration. Putting the acceleration and the mass of the body in the equation of motion provides a mathematical description of the force causing the motion.

For even those not well versed in mathematics, the results of Newton's discoveries are really quite simple, and elegant. The physical world consists of material bodies in motion. Matter possesses the property of mass, which we can measure quantitatively. Material bodies exert forces on other material bodies. The mass of an object and the force acting on it completely determine how it moves, described mathematically by Newton's equation of motion. With that our mathematical description of motion in principle is complete, except for one thing. We might expect to find some connection between mass and the forces that bodies exert on one another. The equation of motion, $F = ma$, may appear at first sight to give us such a connection, but it only specifies a relation between force and the mass of the body on which the force *acts*. What we want is a connection between force and the mass of the body *causing* the force. There is such a connection, in the gravitational attraction between two bodies, and, not surprisingly, Newton also discovered it.

n at a Distance

German astronomer and mathematician Johannes Kepler studied Tycho Brahe's astronomical data and concluded that the planets move in elliptical orbits around the sun. From Newton's laws of motion, some force must be causing them to do that; otherwise they would move in straight paths. Instead, the motion of the planet is continuously bent toward the sun by a force of attraction between the two bodies. From the equation of motion, we can determine—as Newton did using the calculus—that this force of attraction has to vary in magnitude as the inverse square of the distance between the two bodies. In other words, the attraction increases as the distance between the bodies decreases, and vice versa. The inverse square dependence is mathematically necessary in the equation of motion in order for the orbit to be an ellipse.

Newton also assumed that there was nothing special about the sun and the planets, that this force of attraction must be a universal property of matter, and that it should exist between any two material bodies by virtue of their respective masses. The force that holds the moon in its orbit is the same one that makes the apple fall. Galileo had discovered that all bodies at the surface of the earth fall with the same acceleration. Newton actually compared the acceleration of the moon in its free fall around the earth with that of a falling object on the earth and found that the values of these accelerations were indeed in the same ratio as the inverse square of their respective distances from the center of the earth, confirming in his mind the universal nature of the gravitational force. In Newton's own words: "I deduced that the forces which keep the planets in their orbs must be reciprocally as the squares of their distances from the centers about which they revolve; and thereby compared the force requisite to keep the moon in her orb with [the] force of gravity at the surface of the earth; and found them answer pretty nearly."

From the second law of motion, we know that the gravitational force on a body has to be proportional to its mass. Otherwise its acceleration, the ratio of force to mass, could not be independent of mass as it is for bodies in free fall at the earth's surface. From the third law, we know also that the gravitational attraction between two bodies exerts an equal (and opposite) force on each one. Hence the gravitational force between two bodies must be proportional to the mass of each one, in other words, proportional to the product of their masses. We have now taken into account everything we need to know from the laws of motion and from observations to be able to state Newton's law of universal gravitation and to write down the exact mathematical form taken by the gravitational force. In words: any two bodies—that we will label 1 and 2—in the universe attract each other with a force that is directly proportional to the product of their masses and inversely proportional to the square of the distance between them; or, in symbols,

$$F = \frac{Gm_1m_2}{r_{12}^2},$$

where m_1 and m_2 are the masses respectively of the two bodies, and r_{12} is the distance between them.

In this expression, G is a universal constant of nature, the magnitude of which determines the magnitude of the force for two bodies of a given mass separated by a given distance. The value of G has to be determined experimentally by actually measuring the force of attraction between two bodies of known mass a specified distance apart. G is a very small quantity, having a numerical value of about 6.67×10^{-11} (when mass is measured in kilograms, distance in meters, and force in newtons) corresponding to the exceedingly weak nature of gravity. (That is a decimal followed by ten zeros and 667, a number less than one ten-billionth.) As you might imagine, since G is so small, it is also exceedingly difficult to determine its value with any degree of accuracy, and, after more than three centuries of efforts to improve the measurement of G, it is still the least well known of any of the fundamental constants of nature. In spite of the central role that it plays in our lives, it is rather curious and surprising to learn that gravity is the weakest force by far found in nature.

We can get a sense of that by the relative ease with which two bodies, even quite massive ones, can be separated. We do not notice any force due to the gravitational attraction between them, but only their weight, and that only because of the overwhelming mass of the earth.

We need to clear up an ambiguity in what we mean by the distance between two bodies in our statement of the law of gravity. As we have stated it, the law and its mathematical expression apply only to *particles* of mass—material bodies of negligible size. A particle is one of those idealizations, or abstractions, used by the physicist. In the limit of diminishing size, a particle becomes a point mass—literally a material point, a body possessing mass but no physical extent. In reality there is no such thing as a point mass, but it is a useful concept in those situations where we can safely ignore the size of a body and replace it with an equivalent mass located at some point in space. An extended material body can be divided into more and more ever-smaller and smaller pieces, or particles, and between any two such particles the law of gravity applies. The gravitational force of a body on itself is zero, since the forces between the individual particles making up the body cancel out in equal and opposite pairs, by the third law. If the gravitational force of a body on itself were not zero, then the body would spontaneously accelerate, by the second law. To find the gravitational force between two extended bodies, we have to add up the forces between each particle in one body and all of the particles in the other body, using the techniques of integral calculus discussed in chapter 7. When we actually carry out this calculation, we discover that an extended body has the same force on it as an equivalent amount of mass located at a point called the *center of gravity* of the body. For example, a uniform solid sphere or spherical shell can be replaced by an equivalent point mass located at its geometric center. As far as the gravitational force is concerned, a sphere acts like an equal amount of mass located at the center of the sphere.

A uniform solid cylinder or cylindrical shell can be replaced by an equivalent point mass located along the cylindrical axis of symmetry midway between the ends of the cylinder. For any uniform body possessing symmetry—such as a cube or a sphere—the center of gravity is located at the center of symmetry, that is at the intersection of any axes or planes of symmetry. For bodies without symmetry, or for nonuniform distributions of mass, the location of the center of gravity has to be determined using the calculus or measured experimentally. The details of how to do that need not concern us here. All we need to know is that any extended body acts like the same quantity of mass located at its center of gravity. Working out the mathematical proof of this crucial result was one of the significant loose ends that Newton had to tie up before finally publishing the *Principia*.

Now we can give an unambiguous meaning to the distance between two bodies in our statement of the law of gravity: it is just the distance between their respective centers of gravity. For the earth orbiting the sun, or the moon orbiting the earth, it is the distance between the center of the earth and the center of the sun, or the center of the moon, if we treat each of these bodies as a uniform sphere. For a body on the surface of the earth, it is effectively the radius of the earth, since the size of any body on the earth, even the tallest mountain, is negligible compared with the radius of the earth itself. In any case it is always the distance between two distinct points.

As if his other accomplishments were not considerable enough, Newton's crowning achievement was deducing the principle of universal gravitation and discovering the correct mathematical form for the gravitational force. This achievement brought together his other accomplishments and unified them into a complete and consistent understanding of the motion of material bodies under the influence of mass. It would be difficult to say which is the most impressive feature of Newton's scheme, its scope or its simplicity. Substituting the expression for the gravitational force into the equation of motion gives a single, universal description of all gravitational motion. The gravitational force depends only on the masses of the interacting bodies and the distance between them. The right-hand side of the equation of motion depends only on the mass and acceleration of the body whose motion it describes. The equation of motion, along with the initial speed and direction of a body, are sufficient to determine its path and velocity at all subsequent times. It is a grand and elegant synthesis. And it is the standard by which all subsequent mathematical physics would be measured.

Here was the blueprint of the complete universe, the grand plan of the Creator by which everything in heaven and earth moved, all wrapped up in one neat little equation, succinctly stated in a single mathematical sentence that could be parsed and diagrammed and fully understood by mathematicians. The universe was simple after all, a mechanism not unlike a clock in its orderly predictable motions, neatly programmed by the equation of motion and the force of gravity, set into motion by a Creator who was variously the consummate clock maker or the ultimate mathematical physicist. And Newton was the fortunate individual on whom divine Providence had smiled by allowing him to discern how it all worked. The poet Alexander Pope said of Newton, as his epitaph, "Nature and Nature's laws lay hid in night. God said, 'Let Newton be!' and all was light."

Suddenly gone were all the Aristotelian objections to the motion of the earth. The earth and everything on it had been set into motion together and would continue moving in unison until some outside force intervened, in accordance with the first and second laws. Forces between the earth and objects on it were gravitational and were directed toward the center of the earth. An object on the earth's surface moved from west to east with the earth as it rotated about its axis, even when it lost contact with the earth by being dropped or propelled upward, because no other force acted to impede its motion in that direction. Material bodies did not fall or rise to attain their rightful place in the universe but moved as a result of the forces of gravity and buoyancy. Moving objects came to rest not because that was the natural state of motion but because external forces, like the force of friction, opposed their motion with respect to the earth and brought them back to the speed with which the earth and everything on it rotated about its axis. Aristotle was wrong and Newton was right. The king was dead; long live the king.

Newton's description not only made sense out of what was observed but also provided powerful new insights into the nature of motion. For an object on the earth, the equation of motion, with the gravitational force inserted, takes the form

$$\frac{GMm}{r^2} = mg,$$

where M is the mass of the earth, m is the mass of the object, r is the distance of the object from the center of the earth, and g is the acceleration caused by gravity. The first thing we note is that the mass of the object appears on both sides of the equation, on the left side in the expression for the gravitational force and on the right side in its usual place in Newton's equation expressing the second law of motion. Consequently we can divide both sides of the equation by m to obtain the expression

$$\frac{GM}{r^2} = g,$$

from which we see that the acceleration of a falling body is independent of its mass (or its weight) just as Galileo had discovered. Since all points on the earth's surface are very nearly the same distance from its center, given by the average radius of the earth, we can also see why the acceleration due to gravity is approximately constant as Galileo found in his experiments. Substituting the earth's radius, R, into the last expression gives the value of g as

$$\frac{GM}{R^2} = g.$$

Once we have measured the values of g and G and the earth's radius, which had been known approximately since the time of the ancient Greeks, we can calculate the mass of the earth from this expression as $M = \frac{gR^2}{G}$. Newton became the first man to, in essence, "weigh" the earth. Similarly, in case after case the equation of motion gave results in agreement with the observed motion of bodies on the earth and made those observations understandable in terms of the laws of motion and the force of gravity.

But the supreme triumph was in being able to mathematically derive Kepler's laws of planetary motion. The results that had cost Kepler several decades of struggle could be deduced by Newton in the work of a single evening as a mathematical consequence of the equation of motion and the gravitational force of attraction between the sun and the planets. In addition Newton discovered the underlying simplicity unifying all celestial motions that Kepler—and Aristotle before him—had sought in the assumed perfection of circular orbits, but that had, Kepler believed, finally eluded him when he discovered that the planetary orbits were not composed of circles after all but were elliptical paths around the sun. Newton's equation of motion showed that, for a body moving under the influence of the gravitational force, the path of the motion could be any of the *conic sections*, depending on the particular values of position and velocity assumed by the moving body. Ironically, the conic sections were well known to the Greeks, who by the time of Aristotle were the mathematical masters of their understanding. The conic sections are the curves produced by the intersection of a plane with the surface of a right circular cone (a cone in which the base is a circle, forming a right angle with the axis of the cone), as the angle between the plane and the axis of the cone takes on all values. The conic sections consist of: a point (when the plane is just tangent to the apex of the cone); a straight line (when the plane is tangent to the conical surface); a circle (when the plane is perpendicular to the axis of the cone); and an ellipse, a parabola, or a hyperbola (as the angle between the plane and the axis of the cone takes on other values).

All of the conic sections, not just the ellipse, are possible paths for a body moving under the influence of gravity. This is the final unity that the Greeks were searching for in their understanding of motion. In accordance with the first law, a point or a straight line represents the path of two bodies infinitely far apart so that the gravitational attraction between them vanishes. The planets move in elliptical paths, of which the circle is merely a special case occurring when the minor and major axes of the ellipses are equal. The earth's orbit is nearly circular, as is the orbit of the moon around the earth. An earth satellite can be placed in a circular orbit or in an elongated elliptical orbit by adjusting the speed and angle of the trajectory. A projectile on the earth has a parabolic trajectory, as Galileo determined. Comets passing through the solar system can follow hyperbolic paths.

The Greeks singled out the circle as special because of its perfect symmetry. But in doing so, they overlooked an even more fundamental feature of the circle: that it is merely one of a family of plane curves all generated in the same basic manner, by the intersection of a plane and the surface of a right circular cone. What is truly special about these curves are not the differences in their symmetry and geometric appearance, but the common property they all share as conic sections. We can show for instance that the conic sections are those curves traced out by a point moving so that the ratio of its distance from a fixed point to its distance from a fixed line remains constant. The value of this ratio, called the *eccentricity*, determines which of the conic sections the moving point traces. If we let the path of this moving point be described by its position along a horizontal axis (denoted by x) and its position along a vertical axis (denoted by y), then all of the conic sections can be described in a single mathematical sentence by the expression

$$Ax^2 + By^2 + Cxy + Dx + Ey + F = 0,$$

in which A, B, C, D, E, and F are numerical constants. The specific values of these constants determine which conic section the equation describes. Kepler expressed dismay that the symmetry and perfection of the circle was not what unified celestial motions. Newton discovered a far deeper and more powerful symmetry and perfection, one that unified not just celestial motions but all gravitational motion. This was the true symmetry and perfection that the ancient Greeks had been searching for all along, one that identified and singled out what was universal—the Platonic Form—and common to all motion, celestial and earthly. The answer was right under their noses the whole time, in the very geometric properties that they themselves did more to discover and understand than anyone, but they were not properly prepared to look for it. That it took a mathematical form when it was eventually discovered would have been particularly reassuring to Plato, who considered mathematics the proper model for all knowledge. This intellectual quest to discover ever-deeper and more encompassing unity in our mathematical description of nature has been one of the constant themes in constructing the physicist's world, from the beginning to the present moment, elevated at times to a meta-principle with the status of being almost the fundamental principle of the inquiry.

One such unifying concept is that of energy, which also comes directly out of Newton's equation of motion. Energy, like momentum, measures the capacity of a material body to set other bodies into motion or to change the state of their motion. Both momentum and energy are related to the force acting on a body, but in different ways. The rate at which the momentum of a body changes is a direct measure of the force acting on it, expressed by Newton's equation of motion. A force acting on a body for a given length of time will impart momentum to it. The momentum of a body then is a measure of the force acting on it through a specified interval of time. Energy, however, is a measure of the force acting on a body through a given interval of distance. Force acting through a distance is a quantity that the physicist calls the *work* done by the force.

To see how this energy principle arises quite naturally from the laws of motion, we start with the equation of motion, $F = ma$, for a body of constant mass m acted upon by an applied force F producing an acceleration a. Over an arbitrarily short period of time, δt, we can express the acceleration of the body as the change, δv, in its velocity divided by the length of the interval, or

$$a = \frac{\delta v}{\delta t}.$$

With this expression, the equation of motion takes on the form

$$F = m\frac{\delta v}{\delta t}.$$

Upon multiplying both sides of this equation by the product $v\delta t$, we obtain

$$Fv\delta t = mv\delta v.$$

But over the same arbitrarily short interval of time the distance, δx, the body moves is just its velocity multiplied by the length of the time interval, or $\delta x = v\delta t$, and using this expression in the last equation gives us the result

$$F\delta x = mv\delta v.$$

Now we can determine that the right side of this last expression is approximately the same as $\delta\left(\frac{mv^2}{2}\right)$ by simply evaluating the change in the quantity $\frac{mv^2}{2}$, as v changes by an arbitrarily small amount, from say v_1 to v_2:

$$\delta\left(\frac{mv^2}{2}\right) = \frac{1}{2}m\left(v_2^2 - v_1^2\right) = \frac{1}{2}m(v_2 - v_1)(v_2 + v_1).$$

But $v_2 - v_1$ is just the change in v, or $v_2 - v_1 = \delta v$; and $v_1 + v_2$ is approximately equal to $2v$, where v is a value somewhere between v_1 and v_2. Thus

$$\delta\left(\frac{mv^2}{2}\right) = mv\delta v$$

as claimed. Using this result changes the equation of motion to the form that we want, namely

$$F\delta x = \delta\left(\frac{mv^2}{2}\right).$$

This last expression has a particularly meaningful interpretation. It says that there is a quantity, $\frac{mv^2}{2}$, whose change during any arbitrarily short interval of time is given by the force acting on the body during that interval multiplied by the distance the body moves in the same interval of time. In order for this relationship to hold exactly, we must let the size of the time interval approach zero (so that $v_1 + v_2$ will approach closer and closer to $2v$), but this we are free to do by dividing any finite length of time into increasingly many intervals of ever-decreasing duration as we did in our previous discussions of the calculus.

We call the quantity $\frac{mv^2}{2}$ the kinetic, or motional, energy of the body. We define the product $F\delta x$ on the left side of the equation to be what we shall mean by the work done by the applied force F in moving the body a distance δx. If the force happens to be constant everywhere, then the work done is just the product of the force and the total distance through which the body is moved. If the force is not constant everywhere, then we have to evaluate the product $F\delta x$ for each interval of distance that the body moves and add up the results for all the intervals to obtain the total work done, using the techniques of the integral calculus.

With this definition of what we mean by work, the equation of motion says that the change in the kinetic energy of a body during any interval of its motion is equal to the work done by the force acting on it during that same interval. This is the energy principle that we set out to discover. As you can see from the way we obtained it, it is nothing more than the equation of motion in a modified form, and as such it is not a new or separate result. But it provides us with an additional insight into the meaning of the equation of motion.

Forces possess the capacity to accelerate material bodies, either to speed them up or to slow them down. Two bodies moving toward each other under the influence of gravity, such as a body held above the earth and released, will speed up since the force is in the direction of motion. Two bodies moving apart, such as a projectile fired upward from the surface of the earth, will slow down since the force opposes the direction of motion. A body held above the earth is said to possess *potential*, or *positional*, energy by virtue of its separation from the earth. A certain amount of work, equal to the weight of the body times its height above the earth, had to be done on it against the constant force of gravity to elevate it to its position. When the body falls back to earth, the force of gravity will do the same amount of work on the body to accelerate it and increase its kinetic energy. As it falls, the body will lose potential energy and gain an equal amount of kinetic energy. When it strikes the earth, the body will have lost all of its potential energy, as the force of gravity can no longer move it toward the earth, and at the time of impact its energy will be all kinetic. A projectile fired straight upward starts out with a certain amount of kinetic energy but no potential energy. As it rises it gains potential energy and loses an equal amount of kinetic energy. It will continue to rise until it reaches a height at which its potential energy is equal to its initial kinetic energy, before falling back to earth and reversing the process. The energy principle tells us that the increase in kinetic energy is equal to the decrease in potential energy, and vice versa, so that the sum of kinetic and potential energy, which we call the total energy of the body, stays constant. Another meaning to the equation of motion then is that bodies move under the influence of gravity in such a way that their total energy remains constant.

In reading what was just said about energy, you may have thought of an interesting question associated with the example of a falling object striking the earth. If the object moves so that its total energy remains constant, then what happens to the kinetic energy it had just before it collided with the earth and came to a stop? The answer is that work must be done on the object to stop its motion. This work is provided by the contact forces between the falling object and whatever it happens to collide with. The distance through which these contact forces act is the deformation produced in the moving object and anything it hits. An amount of work equal to its kinetic energy at impact is done to change the shape of the objects involved in the collision or even to tear them apart and impart motion to the pieces. But what has become of the energy when everything is finally at rest? Later we will find that the energy principle is universal. Energy it turns out is never created or destroyed. It can only be transferred from one form to another, such as from kinetic energy to potential energy. This principle will lead us to identify heat as a form of energy associated with the internal motions of material substances. The energy of the colliding bodies is ultimately dissipated as heat.

Newton's equation of motion, along with the law of universal gravitation, solved the problem of the motion of the planets and the problem of inaccuracies of the calendar. Now accurate computations of the earth's motion around the sun and length of the year could be made. Even subtle effects such as the gravitational tug of one planet on another as they passed nearby in their orbits around the sun could be taken into account. The heliocentric view of planetary motion was confirmed again and again in each new computation. By the end of the seventeenth century, no serious astronomer any longer doubted the heliocentric model of the solar system. Though it would take the Church several centuries to publicly acknowledge it, the Copernican scheme and Galileo had been vindicated even before the close of the century in which he died. "Eppur si muove"; the earth does in fact move.

Yet for all its success Newton's depiction of gravity was something of a scandal. How could a force possibly reach across the vast emptiness of space to pull two distant bodies together? By what conceivable mechanism could force be transmitted through space even if that space were not empty but only filled with a tenuous medium like the supposed aether that allowed other bodies to move freely through it? This *action at a distance*, as it came to be called, seemed on the face of it absurd to many. Newton's critics wanted evidence of some mechanism by which a force could be transmitted between two distant bodies. It was easy enough to believe in the existence of forces when two bodies collided or when one of them pushed or pulled on the other by direct contact. But between two objects not in contact, there was nothing physical or material by which the force could be imagined to link one to the other. And action at a distance was assumed to affect motion instantaneously, the way two bodies did when they collided or made direct contact. The equation for gravity does not contain any provision for the time required to transmit the force from one body to another. Time does not enter the equation at all. Any change in the motion of one body immediately changes the motion of the other. Newton's law of gravity represents *instantaneous action at a distance*, making it even more puzzling.

Newton replied to these criticisms by confessing that he could not conceive of any mechanical explanation, but that it didn't matter. All that mattered was that he could describe how material bodies move by postulating the existence of a universal force between any two bodies, proportional to the mass of each and inversely proportional to the square of the distance between them. That description of the motion agreed in all respects with what was actually observed, from which Newton concluded that the force must exist and behave as he had described it. What was important here again was not *why* the force behaved as it does, but *how* it behaves, represented by a mathematical description.

The only task of physics, we said, is to describe mathematically the motion of material bodies. The question that arises is whether the components of that description are themselves physically real or merely intellectual concepts useful to the description. What seems undeniably real are material bodies in motion. Matter exhibits the property of mass or inertia, the tendency to resist changes in its motion. Yet we observe changes in the motion of bodies when they appear to come into contact, as in collisions, and even when they are not in contact, as in the case of a body in free fall. We postulate the existence of forces in both cases to account for these observed changes. Forces are non-material quantities, but they are somehow the result of interactions between material bodies. In that sense they are a material property. But they are also a property of space—the gravitational force depends on the distance between two bodies—and later we will discover that they are dependent on time as well.

Since forces are not material quantities, are they real in the sense that matter seems real to us? In our discussion of the second law, we adopted an operational definition for the mass of a body, by specifying a set of physical steps by which the mass of any body could be measured relative to some other body chosen as a universal standard. We then defined the force acting on a body as the product of its mass and its acceleration. This approach has the effect of making mass look like the fundamental quantity, while force appears to be merely a name that we give to the product of mass and acceleration. That's why it was pointed out that we could have proceeded in the opposite fashion, by adopting an operational definition of force as a push or a pull and then using that definition to define mass as the ratio of the force acting on a body to its acceleration. In that approach, force plays the fundamental role and mass is just the name that we apply to the ratio of force and acceleration.

It is important to keep in mind that mass is not an absolute concept. It too is a material property, merely describing how the motion of a body changes relative to the change in motion of any other body with which it interacts. Mass only seems real and substantial because we can reach out and touch material objects and hold them in our hands and directly experience their substance as a resistance to changes in motion, as when we lift an object having mass or try to speed up or slow down its motion. Force, in contrast, seems more like a description of what we experience rather than something physically real in the sense that mass is real. But that impression is an illusion created partly by assigning measured values to mass, even though the numbers are completely relative and arbitrary and represent nothing absolute any more than do the measured values of force. The transmission of forces across the spaces intervening between material objects adds to the impression of unreality, but that too is an illusion.

The physicist is just as hard-pressed as the philosopher or the average person to say where in a material object the property of mass resides, or of what exactly it consists, any more than Newton could satisfy his critics about a mechanism for action at a distance. If mass seems more real because material bodies are tangible, we must keep in mind that the presence of one material body can produce changes in the motion of another body far removed and that these changes can be measured and are no less real or tangible. Both force and mass are numbers determined by a set of operations by which they are measured, and therein resides their only meaning and whatever reality they represent. To the question: "Yes, but what are they really?" there is no other answer than the one provided by the operational definition of their meaning. What is real are material bodies in motion. As for force and mass, each is only as real as the other.

For all the controversy surrounding it, the central issue in Newton's scheme of describing motion was not action at a distance. The real issue was Newton's use of space and time as absolute concepts. In order to apply the laws of motion or even to make sense of them, we have to be able to say what is moving and what is at rest. The equation of motion, which summarizes the essential content of the three laws of motion, requires that we have an unambiguous way of representing the absolute acceleration of a body, which means a way of measuring its absolute position as a function of time. For that we have to be able to identify the fixed and immovable space through which all material bodies are assumed to move. We don't have to require that the frame of reference we use to measure position be absolutely at rest, only that it be moving uniformly through fixed space. Two observers, one absolutely at rest and the other in uniform motion (moving at constant speed in a straight line) will both measure the same value for the acceleration of any other body, since the two observers are not accelerating relative to each another. As a result, they will both conclude that the same force is acting on the body. Therefore, the equation of motion will take exactly the same mathematical form for these two observers. The laws of motion then make it impossible to distinguish between a frame of reference at rest and one moving uniformly, and our description of motion will be unambiguous if we measure position and acceleration using either.

But that doesn't let us off the hook or even lessen the task. We still have to be able to specify what is at rest in the universe in order to decide if a reference frame is moving uniformly. This is the most difficult question posed by Newton's description of motion, one for which he provided no satisfactory answer. There seems to be no absolute and independent means of determining what, if anything, is at rest. The best we can do is to measure relative motion—the motion of one body relative to others. Without some way of actually determining a frame of reference with respect to which everything else in the universe can be shown to be moving, the concept of absolute space is one that cannot be made convincing. We may wish to believe the concept is valid on philosophical or other grounds, but our mathematical description demands that we be able to give a strictly operational definition, in terms of how we can actually measure it, to the concept of absolute motion. Failing that, our entire scheme is suspect.

Why then did Newton insist that space is absolute? Newton's answer was that we can use the forces acting on a body to demonstrate its absolute motion. The acceleration indicated in the equation of motion, according to Newton, is relative to the fixed framework of space itself and is thus absolute. If a body is at rest or moving uniformly, there can be no forces acting on it, and we will not be able to detect its absolute motion. But if the body is accelerating, there will be a force acting on it, and we can observe the effects of this force to confirm that the body is undergoing absolute acceleration, which means absolute acceleration with respect to fixed space.

Newton's famous example of this is a pail of water suspended by its weight from a twisted rope. If the pail is at rest or moving uniformly, all parts of the pail and its contents exhibit the same uniform motion and the surface of the water remains flat and undisturbed. As the rope untwists and the pail spins about its axis, all parts of the water want to obey the first law and move in straight paths but are constrained by forces exerted by the sides of the pail to move in circular paths around the axis of rotation. The acceleration of the water creates a force that pushes the water outward from the center so that the surface of the water takes on a concave shape. The concave surface tells us that the water is accelerating, even if we were able to make no other observations to confirm the rotational motion of the pail. According to Newton, we could imagine that the pail of water is the only object in the fixed space of an otherwise empty universe. Whether the pail is rotating or not could still be ascertained by the action of the water. Similarly, the forces acting on any body or between the individual parts of the body can be used to detect its absolute acceleration. Two objects connected together by a spring must be rotating if they cause the spring to stretch. Later this same argument was used by Albert Einstein to construct a completely new—and totally different—description of gravity.

Forces then, according to Newton, are the test of absolute motion. Two observers who appear to measure different forces acting on the same body can be certain that at least one of them is accelerating. This has a very practical implication for us, living as we do on the surface of the earth. As the earth rotates around its axis and travels in its elliptical orbit around the sun, everything on it is accelerating through space. Objects on the earth should move as if they are being acted upon by forces in addition to the simple force of gravity. These other apparent forces produced by the earth's acceleration can be measured, and their effects are easily observable. One of the most evident is the rotation produced in the earth's atmosphere as it moves between regions of differing pressure, causing it to swirl counterclockwise in the northern hemisphere around a region of low pressure and clockwise around a region of high pressure. Much of the earth's weather phenomena can be traced to this effect. As another example, we can observe the slow rotation of the plane of oscillation exhibited by a long pendulum, known as a Foucault pendulum after its discoverer, Léon Foucault. An observer on earth might conclude that these motions are caused by real forces acting in addition to the force of gravity. But an observer stationed in space and not undergoing acceleration would see them only as the gravitational force combined with the various accelerated motions of the earth.

Though it may seem even less questionable on philosophical grounds than the idea of absolute space, Newton's concept of an absolute time that "of itself and from its own nature, flows equably without relation to anything external" is equally problematic on operational grounds. Absolute time seems like a good enough idea until one begins to question how it might be confirmed by measuring it. Distance, we said, is what we measure with a ruler or a meter stick and, operationally, time is what we measure with a clock. If we wish to confirm even something as simple as whether a particular interval of time—which was what Newton termed relative time—measured in one place is the same as a supposedly equal interval measured somewhere else, then we have to compare identical clocks located at different places. The issue becomes whether they all run the same, which is really asking whether physical processes are identical at all places in the universe. To compare all of these clocks, we would need a way to read them all at the same instant. That would be straightforward if they were all located at the same place. It is the fact that they are not, that complicates things. Since they are physically separated, we either have to transmit readings through space with no delay (meaning at infinite speed) or have some way of unambiguously correcting for whatever time is required to transmit the readings from one place to another. Correcting for the transmission delay means that we have to be able to measure the distance absolutely. But we so far have no way of doing that. And what if the clocks themselves are moving? Does that have any effect on how they measure time? Since we cannot determine whether an object is at rest or moving uniformly, we have to ask how motion may affect the operation of our clocks.

These are all questions that went largely ignored until the beginning of the twentieth century, mostly because until then, for reasons that will be discussed later, there was no empirical reason to doubt the universality of time. Debate about absolute time seemed best left to philosophy and philosophers. Physics dealt with the relative time measured by clocks and mostly with local time at that. But a strictly operational meaning to time demands that all of the operations by which the values of time are to be measured and compared be carefully spelled out and that the results be unequivocal. And, from what we have already said, it is clear that the concepts of absolute time and absolute space will in some way be connected.

Newton took as the frame of reference for absolute space the so-called fixed stars. They do not appear to be moving relative to the other stars. Using them as our reference, the motions of the planets and other bodies in our solar system can be consistently described. Even the apparent motion of the sun with respect to the fixed stars appears to be largely accounted for by the motion of the earth in its orbit around the sun, and the sun itself serves as an approximate fixed reference frame that we can use to measure position and motion in our mathematical description. With the fixed stars as our reference, the Newtonian description appears to work, and that is the ultimate test and final vindication of Newton's choice. It does seem that there is a general background of space, represented by the region occupied by the most distant stars, through which the absolute motion of material bodies makes sense, at least locally, as a concept. We have no way of demonstrating that this background is absolutely at rest or moving uniformly, or even whether such a question makes any sense, but it does seem to serve as an appropriate framework against which to view the motion of bodies nearby.

It was against the backdrop of overwhelming success enjoyed by the Newtonian scheme that David Hume questioned another of the absolute concepts of Newton's description of motion—and of all science—that of causality. Leave it to an English empiricist philosopher to raise such a mundane and practical question of how we can be certain that two events always observed to occur together in the past, such as the collision of two bodies and their subsequent recoil, will continue to be associated in the future. We say that the collision causes the recoil because we have always observed them to occur in the same sequence in the past. What assurance, Hume asked, do we have that we will observe the same sequence of events on the very next try? Nothing but habit, or custom, Hume answered; in the process he voiced a fundamental skepticism about the logical foundation of all science. Causality is not absolute, Hume argued, but is conditioned by experience the same way that our concepts of space and time are conditioned by experience.

Immanuel Kant simply did not want to believe that Hume could be right. Aroused from what he termed his "dogmatic slumbers" by Hume's skepticism, Kant mounted a counterargument, resulting in his monumental *Critique of Pure Reason*, a work so intractable and difficult that at the urging of others he published a summary of his argument entitled *Prolegomena to Any Future Metaphysics* intended for a wider audience. Kant argued that causality is not conditioned by, but is a condition of, all experience. We do not derive the concept of causality from our experience, he said; rather our experience is made possible because of the a priori existence of causality. That is the gist of Kant's argument. The success of Newtonian physics, he said, demonstrates that the universe obeys laws. Experience is required in order to determine the specific content of the laws; nothing prior to experience can establish that; but the form of the laws—even the very possibility of natural laws—is based on certain a priori concepts or categories without which experience itself would be impossible. Among them are space, time, and causality. These must exist a priori as necessary conditions for our experience and the judgments we make about that experience through human reason. Kant in his argument sides with Plato—not surprising for a philosopher who believed that metaphysics was possible—that the natural laws and any preconditions necessary for their understanding are discovered truths about the physical world, rather than creations of the human intellect or mental constructs arrived at to organize and make sense out of what we experience.

If Kant had lived a century later and had encountered Darwin's theory of evolution, he might have constructed a more cogent argument. The universe is perhaps spatial, temporal, and causal. And our minds, having evolved in the world as products of its physical processes, are perhaps hard-wired in accordance with the concepts of spatiality, temporality, and causality. Then of course these concepts would be a priori to human reason, because they would be part of the structure by which we reason. They would be the basic physical principles by which reason and the human mind operates, much as Kant argued, albeit for somewhat different reasons. After two centuries, it is still a matter of debate whether Kant said anything at all in his answer to Hume, or whether he merely asserted at great length what he had set out in the beginning to demonstrate. Whether the laws of motion confirm causality, or merely express it, is still arguable. For most, Hume's skepticism has gone unanswered, simply because it is unanswerable.

The laws of nature exist (they seem to describe the working of the world), but there is no demonstrated necessity associated with them, either in their existence or in their specific form. They are based on concepts—like space, time, and causality—that are not demonstrably absolute but are merely provisional and subject to experience, not a priori to it. Certainly no one seriously doubts the validity of causality as a principle of the natural world, or expects to see any violations of those natural laws based on its assumption. Very likely Kant is correct in his assertion. The issue is whether one can demonstrate causality to be a necessity a priori to experience, or whether it is only derived from our experience and hence can never be established absolutely in any independent way. All we can ever know with certainty about the world is what we experience. We may draw conclusions based on our experience, but those conclusions are always subject to verification by further experience. According to this view, it is impossible to demonstrate the logical necessity of causality any more than it would be possible to demonstrate the logical necessity of three dimensions for the world, or four or six or eleven. Demonstrating the absolute nature of causality takes its place alongside demonstrating the absolute motion of a body as one of the fundamental limits to human knowledge.

The other thing required for a complete description of motion—in addition to some way of specifying position absolutely—is a separate and independent description of the forces acting on a body. If the equation of motion is not to be taken as merely a definition of what we mean by a force but a genuine mathematical relationship between the force acting on a body and its acceleration from which the motion of the body can be determined once the force is known, then we must have an independent description of the force. Newton's law of gravity is the model, specifying how the gravitational force varies with the mass of any two interacting bodies and the distance separating them. Its discovery raises the possibility of other types of forces. Newton clearly regarded gravity as the fundamental force shaping the natural world. Other forces were those causing changes in the momentum of bodies during collisions. Newton gave these contact forces no separate description, independent of the product of mass and acceleration, and was unaware of their fundamental nature. He was aware that they were somehow connected with the property that made matter impenetrable and that held material bodies together, but he associated them with the property of mass and not with some new property of matter.

These other forces, it turns out, are electrical in nature and are far more significant in shaping our immediate experience of the world than gravity, which is of little consequence beyond determining our weight and holding us on the earth. The eventual understanding of them during the two centuries following Newton's *Principia* was to set the stage for the important discoveries of the twentieth century and would eventually force a major revision of both Newton's description of motion and his description of gravity.

10

Matter and Light

Gravity is a universal force of attraction between any two bodies or any two particles of matter, pulling them together. We might wonder then why gravity doesn't cause matter to simply collapse upon itself. The gravitational force increases in strength as two objects come closer together due to the inverse square dependence of the force on the distance of separation. As the distance shrinks to zero the force increases without limit. Why shouldn't the internal forces of attraction continue to shrink a substance until every particle of matter in it is making direct contact on all sides with other particles except at the outer surface? Or continue to shrink a substance until it was a mere point? All matter should be a dense lump, and it should be surprising to find objects with pores or spaces anywhere inside them.

The answer is that there are other powerful forces at work shaping the behavior of materials. These other forces are all due to a single electromagnetic force, one vastly stronger than the gravitational force. They are what make solids impenetrable and liquids viscous and gases tenuous. They are what hold substances together and give them their strength and elasticity. These are the forces acting between bodies when they come in contact and that cause them to recoil in collisions and that lead to friction between objects. The force with which the floor pushes up against the chair I am sitting in is electromagnetic in origin. These other forces shape almost our entire experience of the world through the physical contact we have with everything. They are directly responsible for our five senses and govern all our sensory perceptions. They are the basis of chemistry and all biological life; they are what enable DNA to replicate itself. Electromagnetism not only gives rise to all of the material characteristics of our world, but it is also responsible for the nonmaterial component of the universe that we see as light. By comparison, gravity plays a relatively minor role in shaping our immediate surroundings.

We are usually unaware of the overwhelming strength of electromagnetism, since in most cases these other forces are internal and the net electromagnetic force between two bodies is close to zero. These internal forces play a role only when bodies closely approach each other or actually come into contact. In light of that, we might go back to our original concern and ask why the whole universe doesn't collapse upon itself under the gravitational attraction of matter. One answer might be that the universe is infinite with matter distributed more or less homogeneously throughout space. If that were the case, the gravitational force on any body would be approximately the same in every direction. There could be local clumping due to nonuniformities in the distribution of matter on a local scale, just as we observe with the stars and galaxies, but the whole universe would be kept from collapsing by roughly equal gravitational forces pulling in all directions. Another answer might be that the universe is collapsing, but that it is too young to have completely collapsed yet. The local clumping of matter that we observe might be just the beginning of that process. This raises the question of how the matter ever became dispersed against the force of gravity in the first place. Still another possibility might be that the universe is finite but is expanding so that the kinetic energy of expansion is being expended as work done against the gravitational forces opposing the expansion. Then the question becomes whether the expansion will continue forever or whether it is slowing down and will eventually stop, following which gravity would subsequently cause the universe to collapse. These are the kinds of questions about the motion of matter on a cosmological scale that are raised by our description of motion on a smaller scale and that must be addressed by any theory of cosmology. All observations indicate that the universe is indeed expanding, that the expansion will presumably continue forever, and that in fact is actually speeding up, though the question of the future fate of the universe is at the forefront of research in cosmology at present.

Regardless of the outcome, we can be certain that the universe as a whole can never be in static equilibrium. There is a general mathematical result, sometimes referred to as Earnshaw's theorem, from which we can conclude that it is impossible to find any arrangement of material bodies that will hold them motionless in stable equilibrium under the influence of gravitational or any other kind of forces that vary as the inverse square of the distances between them. Configurations of unstable equilibrium, much like a pencil balanced upright on its point, may exist in which the bodies could in principle remain motionless. But any movement, no matter how slight, would destroy the delicate equilibrium and set the bodies in motion. Only the hand of God could ever be that steady, and even if it had been, at some point he must have flinched. Since the universe is now in motion, it can never again return to a condition of static equilibrium. Earnshaw's theorem is a direct consequence of the inverse square dependence of the gravitational force. Heraclitus's emphasis on motion finds vindication in the mathematical form of Newton's law of gravity. Any arrangement of the universe on a cosmological scale is one necessarily based on movement.

We are considering only the properties of matter required to describe motion. Mass was the first property of that kind that we encountered. The electromagnetic force depends on a second such property of matter, separate and distinct from mass, termed *electrical charge*. The concept of electrical charge is necessary to describe the motion of bodies possessing this property. There seems to be only one kind of mass in the world, and gravity is always a force of attraction. But there are two kinds of electrical charge. We label these two types of charges positive and negative, following a convention first used by Benjamin Franklin. Corresponding to the two kinds of charge there are two kinds of force. Charges of the opposite kind attract one another, but charges of the same kind, whether positive or negative, repel one another.

Ordinarily matter contains identical amounts of positive and negative charge so that the net charge of a body is zero. This is because the total charge of a body is just the sum of its individual charges. But the two kinds of charge reside in different parts of the material. We now know that the negative charge is carried by electrons, which are mobile and can be stripped off a material to give it a net positive charge or added to it to give it a net negative charge. Positive charge is carried by protons, which are much more massive than electrons and are tightly bound up in the atomic nucleus, making them far less mobile than electrons. Electrons can be removed from a material in a whole host of ways, including friction, heat, light, mechanical impact, or a strong electromagnetic force. When a glass rod is rubbed with silk or cat's fur, electrons are transferred from one material to the other, leaving both electrically charged. The material losing electrons acquires a net positive charge and the other material picks up a negative charge of equal magnitude. The same thing can occur as a result of shuffling across a carpet on a dry day, as anyone who has ever received an electrical shock after doing so can attest. In the same manner, friction between wind and earth or between wind and cloud can transfer charge between the clouds and the ground or between different parts of the cloud, producing lightning when the charge subsequently flows between oppositely charged regions.

We can get some idea about the relative strength of the electromagnetic force by comparing the electrical force pushing two electrons apart with the gravitational attraction between them at the same distance of separation. We find that the magnitude of the electrical force is about 10^{42} times that of the gravitational force (a number 1 followed by 42 zeros!). Consequently, we only notice the gravitational force when dealing with bodies of enormous mass like the earth, the moon, and the sun. For bodies on an atomic scale, we can safely ignore gravity compared with the strength of electromagnetic forces. We cannot however ignore mass itself, even on an atomic level. Mass is still the property that determines how electrons and protons respond to all forces, in accordance with the second law of motion. The mass of a proton is about 1,836 times that of an electron, and as a result, the same force acting on both will impart a correspondingly greater acceleration to the electron. Electrons therefore are not only more mobile but in general travel at much greater speeds than protons or more massive objects like atoms and molecules. We see from the enormous strength of the electromagnetic force and its effect on the atomic constituents of matter that we will have to include it in any complete description of motion, particularly in describing the internal motions of material bodies.

The electrical force between two stationary charges (which we term an *electrostatic force*) takes on a familiar mathematical form. The electrostatic force, F_{12}, between two point masses, of charge q_1 and q_2 respectively, situated a distance r_{12} apart, is given by the expression

$$F_{12} = \frac{k_e q_1 q_2}{r_{12}^2},$$

which has exactly the same mathematical form as the gravitational force, with charge taking on the same role here that mass plays in gravity. The electrical constant k_e replaces the gravitational constant G. The letter q is a convention that physicists use to refer to the quantity of electrical charges. We specified point masses so that we would be able to ignore the size of each one in determining how far apart they are. All that is really required are two bodies whose size is negligible compared to their separation. On a macroscopic scale, an electron is a perfectly good point mass; so is a proton or an atom, or even large numbers of atoms. Of course we still have to specify some way of measuring charge in order to attach quantitative values to q_1 and q_2 in this expression. Once we have done that, however, the constant k_e can be determined by measuring the actual force between two given charges located a known distance apart. Due to the strength of the electrostatic force, it is far easier to measure the electrical constant than it is to determine the gravitational constant G, and the numerical value of k_e is found to be 8.988×10^9(10^9 is 1 followed by nine zeros, or one billion). Recall that G is 6.67×10^{-11}, so this constant is twenty orders of magnitude—one sextillion—larger.

Earnshaw's theorem pertains to any force that varies as the inverse square of the distance between bodies. Thus we know immediately that it is not possible to find a stationary configuration of charges that will be in stable equilibrium under the influence of the electrostatic force. Any static arrangement of charges will be unstable, so that even the slightest motion, however infinitesimal, will upset the arrangement and cause the charges to move, never to return to static equilibrium. For charges to be stationary, they must be held in place by something other than an electrostatic force alone. Motion is not only a permanent and inescapable feature of the external world, it is also an essential characteristic of the internal constituents of matter. The only equilibrium possible for matter is a dynamic equilibrium based on the motion of mass and charge. The material objects we see around us, and everything they are made of, must be in motion, whether those motions result from gravity or electromagnetism.

This electrostatic force between any two stationary point charges is known as Coulomb's law in honor of the man who first discovered its behavior, eighteenth-century physicist Charles-Augustin de Coulomb. From the form of Coulomb's law, we see that $F_{12} = F_{21}$, since interchanging q_1

and q_2 does not change the value of the force. This means that the force on q_1 due to the presence of q_2 is the same as the force on q_2 due to the presence of q_1, and the Coulomb force, like the gravitational force, obeys Newton's third law of motion. Coulomb's law says that an electric charge exerts a force everywhere in space on any other charges that happen to be present. And, like the gravitational force, this electrical force represents action at a distance. The Coulomb force due to the presence of charge extends all the way to infinity, even through empty space, just like the gravitational force produced by the presence of mass. Both are long-range forces and go to zero only as the distance of separation becomes infinite.

If instead of point charges, we have an arbitrary distribution of charges in the form of objects larger than a point, or extended bodies, then Coulomb's law no longer takes on the relatively simple form given in the previous expression. The force between extended charged bodies can become extremely complicated to express mathematically, even though it is based on the simple Coulomb force between the individual point charges that make up the extended charge distributions. It becomes more convenient to talk instead about the total force per unit charge acting on a single positive test charge located at any given point in space and to call that quantity the *electric field* at that point. As an operational definition, this means that we place a positive test charge at any point where we wish to know the electric field, measure the total force acting on the test charge, take the ratio of the force to the size of the test charge, and call that ratio the magnitude of the electric field at that point. The direction of the electric field is just the direction of the force exerted on the test charge. We have to be careful to use a test charge small enough not to disturb the charge distribution whose electric field we are trying to measure.

The electric field at any point in space tells us the magnitude and direction of the force exerted on any charge placed there. We simply multiply the value of the charge by the magnitude of the electric field at that point to obtain the force. Electric fields are what cause charges to move, and it is the motion of charged bodies that we are ultimately interested in being able to describe.

Electric fields can be produced in a variety of ways. Anything that causes a separation of charge into negative and positive charge distributions will lead to a Coulomb force between them that represents an electric field. The separation of charge, as we said earlier, can be by means of friction or mechanical work, by heat or light or chemical action, or by any number of other mechanisms. All that is required is some means of supplying the energy necessary to separate positive and negative charges against the Coulomb force of attraction holding them together. A battery, for example, does that chemically and produces an electric field that can be applied to a material called a conductor in which electrons are free to move. When that happens, the electrons move in response to the applied electric field and the resulting flow of charge constitutes an electric current.

We define an electric current quantitatively as the charge per unit time flowing past a given point in a material. If δq represents the change in the quantity of charge at a given location in the material and δt represents the time interval in which that change occurs, then the current at that point, which we will call i, is given by the ratio

$$i = \frac{\delta q}{\delta t}.$$

Electric currents can flow through any kind of medium in which the charge is free to move. Typically currents flow through conducting wires or other metallic paths, in which electrons are mobile and move with less resistance than in other materials but to which the electrons are nevertheless confined, so that the currents can be directed only where we wish them to go. Lightning is an electrical current flowing through the air between clouds or between the clouds and the ground. In the case of lightning, the charge on the clouds and the ground produces an electric field strong enough to pull electrons out of the air molecules, freeing them to flow as current and rendering the air conductive. The earth itself is constantly being bathed by a current of protons from the sun known as the *solar wind*, which flows directly through the intervening space between the sun and earth. There are currents of electrons and other charged particles trapped above the earth in the Van Allen belts by the action of the earth's magnetic field.

When charges move, they travel in the direction of the electric force as required by the laws of motion. An electric current follows the path of the electric force in whatever medium it flows and, as a result, currents trace out lines of electric field in the medium. Moving charges exert a second kind of force distinct from the Coulomb force between stationary charges. This second force, which we refer to as magnetism, exists only when the charges are in motion and acts always at right angles to the direction of motion, meaning at right angles to the flow of current. By the second law of motion, force and acceleration must always be in the same direction. This additional force produces a change in the velocity of the moving charge that is always perpendicular to its velocity. It cannot change the speed of a moving charge but only the direction of its motion. Consequently the magnetic force cannot do any physical work, which as you recall requires a force acting in the direction of motion. The magnetic force has no effect on the speed of a moving charge and, hence, no effect on its kinetic energy. Any physical work that appears to be done by magnetism, such as when an electromagnet picks up a heavy metal object, is actually being done by the electric field that produces the current causing the magnetic force of the electromagnet. Even the work done when a permanent magnet picks up an object is done by the currents of electrons flowing in the atoms that make up the magnetic material. The magnetic force is created by the motion of charge, which in turn is caused by the presence of an electric field, so that the magnetic force, like the Coulomb force, is actually caused by the electric field.

Just as we used the Coulomb force between static charges to define an electric field, we can use the magnetic force between moving charges to define what we mean by the *magnetic field*. We defined the electric field as the force per unit charge acting on a positive test charge placed at any point in space. The magnetic force is between moving charges, or lines of current, so we specify the magnetic field in terms of a force per unit length. Whereas the electric field is a force exerted on electric charge, the magnetic force is exerted on an electric current. So instead of a force per unit charge, we define the magnetic field in terms of a force per unit current. The magnetic field then is the force per unit length per unit current acting on an element of positive current at any point in space.

Expressing this definition in symbols, we have

$$B = \frac{\delta F}{i\delta l},$$

where B represents the magnetic field, and the ratio $\frac{\delta F}{i\delta l}$ represents the force per unit length acting on the current i flowing at any point where we wish to know the magnetic field. It makes perhaps more sense to write this expression in the equivalent form

$$\delta F = Bi\delta l.$$

In this form, it says that, if there is a magnetic field at any point in space, then a length δl of any current i flowing at that point will experience a force δF acting on it. We are using the symbol δl here to represent an arbitrarily small length of current and δF to represent only the force acting on that length of the current. This is consistent with our previous use of δ to represent an arbitrarily small portion of some quantity. It is worth emphasizing that the current in our definition is not the current that produces the magnetic field. It is the current that we use to test for the presence of the field.

This definition gives us the magnitude of B once we know the magnitude of the force per unit length and the value of the current on which it acts. The magnetic field also has a direction and here it differs specifically from the electric field. The electric field is always in the direction of the force acting on a positive charge. The magnetic field is not in the direction of the magnetic force but is always perpendicular to both the force and the direction of the current. The magnetic force is also perpendicular to the direction of the current, so the magnetic field, the force, and the direction of the current are all mutually perpendicular. It is this rather bizarre property that gives the magnetic field its distinctive character and makes its properties much more complicated to deal with than those of the electric field.

The only source of the magnetic force is electric current. When the charges stop moving, the force and, hence, the magnetic field ceases. Magnetism is not produced by any material property that can be isolated to a point in space, like a localized point charge. To produce a magnetic field, the charges have to move and trace out lines of current in space. We say that there are no *point sources* of the magnetic field, nothing analogous to electric charges, and no separate material property is responsible for magnetism. There is no magnetic charge in nature, only electric charge.

Now that we have an operational definition of what is meant by the magnetic field, we have a way of investigating how the magnetic field depends on the currents that produce it and how the field varies with position in space. For example, a wire carrying a current produces a magnetic field. We know that because, if we bring another wire carrying a test current up to the first wire, we find that there is a force between the two wires. By measuring the force per unit length and knowing the value of the test current, we can determine the magnitude and direction of the field. The results of all such experiments lead to a general mathematical description of the magnetic field, known as the Biot-Savart law after the two Frenchmen Jean-Baptiste Biot and Félix Savart who first described it.

The mathematical expression for B is a rather complicated affair in general but in certain cases it reduces to something simpler. For instance, the magnetic field in the region close to a long straight wire carrying a steady current is found to be directly proportional to the current in the wire and inversely proportional to the perpendicular distance from the wire. In symbols,

$$B = 2\frac{k_m i_1}{d},$$

where i_1 is the value of the steady current flowing in the wire and B is the magnitude of the magnetic field a perpendicular distance d from the wire. The magnetic constant k_m in this expression is determined by the size of the units that we use to express B and for our purpose has the value of 10^{-7} (a decimal followed by seven zeros, or one ten-millionth). The small value of k_m relative to k_e means that the magnetic force is much weaker than the electric force. The ratio of these two constants will turn out to be one of the most fundamental of all properties of the universe.

Notice that the magnetic field has the same value for all points the same distance from the wire and depends only on the value of the current flowing through the wire. This means that the magnetic field has the same value along any circle with the wire passing through its center. The direction of the field has to be perpendicular to the direction of the wire (perpendicular to the current), which means that it is tangent to the circle so that the magnetic field curls around the wire. If we place another long straight wire beside the first, with a distance d separating them, this second wire will have a constant magnetic field everywhere along its length. If we then let a current i_2 flow in the second wire, then by our definition of the magnetic field, there will be a force per unit length on the second wire given by

$$\frac{\delta F}{\delta l} = Bi_2 = 2\frac{k_m i_1 i_2}{d},$$

where we have made use of the expression for B from the Biot-Savart law in writing the last expression. If the currents both flow in the same direction, we find that the wires are attracted to one another. If the currents are in opposite directions, the magnetic force pushes them apart.

This last expression gives us a way of defining the unit of current. By measuring the force per unit length between two long straight wires a known distance d apart when the same current i_2 is flowing in both, we can determine the value of the current from this last expression. We have to know k_m, but we know that its value was determined by the units that we use to express the magnetic field. Once we have a way of measuring the value of a steady current, we can find the amount of charge that flows in a specified length of time by simply multiplying the value of the current by the time. We then have an operational way of measuring charge, which up to now we had not specified.

Electric fields produce currents by causing charges to move, and these currents in turn create magnetic fields. There is also a way of having a magnetic field create an electric field. If a conducting material is physically moved through a magnetic field, then the charges in the conductor are like any charges moving in a magnetic field. They experience a force that acts in one direction on the negative charge and in the opposite direction on the positive charge. This magnetic force causes the electrons in the conductor to move, resulting in a separation of positive and negative charge and the creation of an electric field in the material. If the conductor is part of an external circuit, then a current will flow in the circuit as long as the conductor continues to move. Exactly the same effect is produced whether the conductor is moved through the magnetic field or the conductor is fixed and the magnetic field is moved. The electric field created by moving either one depends only on the relative motion between the two. If the direction of the relative motion is reversed, the direction of the electric field and hence the direction of the current is also reversed. By moving the conductor back and forth in the magnetic field, or moving the magnetic field back and forth with respect to the conductor, an alternating electric field and an alternating current can be generated. This is the basic principle by which most of the electrical power in the world today is produced. The motion between conductors and magnetic fields is commonly supplied by water turbines in dams, by steam turbines powered by burning coal, oil, or natural gas, or by heat from a nuclear reactor.

Note that the magnitude of the electric field produced by moving a conductor in a magnetic field depends only on the relative motion between the two. It does not matter whether we consider the conductor to be moving in a stationary magnetic field or the magnetic field to be moving past a stationary conductor. All that matters is the relative motion of one with respect to the other. The principle of special relativity pertains here as it does in the motion of any two bodies. It is not possible by any means—whether mechanical, electrical, or otherwise—to determine the absolute motion of a given body with respect to any other body.

Creating an electric field in this manner is an example of a more general principle known as the *law of induction*, or Faraday's law, after Michael Faraday who described the phenomenon in the nineteenth century. Faraday demonstrated that any change in either the magnitude or the direction of a magnetic field creates, or induces, an associated electric field, even in a vacuum or free space. It is not necessary for there to be any material medium or charge present at all. An electric field always accompanies any change in the magnetic field. The magnitude of the induced electric field depends on the rate of change of the magnetic field. A constant magnetic field, however, does not generate an electric field. Later James Clerk Maxwell showed, mostly on theoretical grounds, that the converse also had to be true. Any variation in the electric field creates an associated magnetic field, and the magnitude of the magnetic field depends in the same way on the rate of change of the electric field.

Using this latter discovery, Maxwell synthesized the laws of electromagnetism much as Newton had done for motion. He showed that all of the properties of electric and magnetic fields could be completely described by four mathematically elegant equations that today bear his name. Maxwell's equations—together with the mathematical expression for the electromagnetic force between charged particles—give us a complete description of how electromagnetism affects the motion of charged bodies and becomes an essential part of our overall description of motion. With these equations and Newton's laws of motion, we can describe the motion of bodies having both mass and charge. But they do even more than that. Maxwell's equations reveal previously unsuspected properties of moving charges that make electric and magnetic fields essential components of the physical world, in the same way that matter is an essential part of the universe.

Writing down Maxwell's four equations in their mathematical form would lead us too far afield. Instead we will merely describe in words what each one says. The first equation expresses the fundamental relationship, satisfied by all electric fields, between the field and electric charge. The equation tells us that the ultimate source of the electric field is charge and that the lines tracing out the direction of the field begin and end on charges, leading from positive charges to negative charges. The lines of force diverge from positive point charges and converge onto negative point charges. The field at every point in space is completely determined by the quantity and spatial configuration of any charge located within the region or on its boundaries. Once we know the magnitude and arrangement of all charges, the electric field is determined.

The second equation has exactly the same mathematical form as the first, with the magnetic field substituted in place of the electric field, and with the term representing charge set equal to zero. It says that there are no localized or point sources of the magnetic field, meaning that there is no "magnetic charge" in nature. The ultimate source of the magnetic field is the same as the electric field, namely electric charge. Since there is no magnetic charge on which the field lines can begin and end, then the field must follow lines in space that are continuous closed curves without beginning or end. The magnetic field lines have no breaks in them and always close upon themselves, like the circular paths tracing out the field direction around a long straight wire carrying current in the example that we used earlier.

The third equation is the Biot-Savart law, as modified by Maxwell, and tells us that magnetic fields are caused by moving electric charges, or currents, and also by fluctuations in the electric field in time. The fourth equation is Faraday's law and has the same form as the third equation, with the electric and magnetic fields interchanged and with the term corresponding to current set equal to zero. This equation tells us that electric fields can also be produced by fluctuations in the magnetic field in time, but that there are no magnetic currents in nature, corresponding to the absence of any magnetic charge.

Maxwell's equations exhibit an elegant and suggestive symmetry. The first two equations, one for the electric field and the other for the magnetic field, have exactly the same form except for the absence of any charge term in the magnetic field equation. In it the value of charge is set equal to zero, corresponding to the absence of a magnetic charge property in nature. The third and fourth equations, each involving both the electric and magnetic fields, also have the same mathematical form with the fields interchanged in the two equations and with the absence of any current of magnetic charge in the equation expressing Faraday's law.

If magnetic charge existed in nature, then Maxwell's equations would become completely symmetric. The electric field would be produced by electric charge and by time fluctuations in the magnetic field (as it is now), but also by the motion of magnetic charge, or "magnetic currents." And the magnetic field would be produced by currents of electric charge and by time fluctuations in the electric field (as it is now), but also by the presence of magnetic charge. What is now merely a mathematically elegant description would be elevated to one nothing short of perfection in its symmetry and completeness. Because in the physicist's world there is a great desire for mathematical symmetry and simplicity, the holes left in Maxwell's equations by the absence of any magnetic charge, or *magnetic monopoles* as they have come to be called, seem to the physicist like some serious blemish in an otherwise stunningly beautiful painting. The rest of the painting makes the blemish all the more glaring and incongruous by contrast. It is almost as if there had been some curious oversight at the moment of creation.

Such is the overwhelming desire for mathematical elegance and cohesiveness in our description of the world that physicists have searched diligently for the missing magnetic monopoles. The conviction is that surely nature would not have been created with such an obvious irregularity. Even if it were, the omission must not really be a flaw but only a hint of some deeper underlying symmetry that will finally unify our understanding, similar to the way Newton's laws of motion did in showing that the gravitational motion of bodies is described not only by circles and ellipses but by all of the conic sections. So far the search for magnetic monopoles has been in vain, other than to place certain limits on their possible properties and abundance. But so compelling is the idea of their existence that the search continues, as it will until there is some ultimate resolution, either by finding the elusive monopoles or, more likely, by finally understanding their absence in terms of some deeper insight. The latter would seem like the more satisfying prospect, since even if they are discovered we will still have to explain why there are so few of them compared to the amount of electric charge in the universe. Even that asymmetry, if it comes to pass, will serve as the departure for continuing to search for other meanings buried in still deeper symmetries and understandings, ad infinitum.

Maxwell's equations allow us to reduce all electromagnetism to these four fundamental relationships between the electric and magnetic fields and electric charge. They are a set of equations, and as such they must be mathematically consistent with one another. All four equations must be true simultaneously, both in form and content. Any mathematician can explore the consequences of that requirement by examining the relationships that exist between the equations themselves. In any region of space where there is no charge, and hence no matter, we can set the charge and current terms in Maxwell's equations equal to zero to get four equations involving only two quantities: the electric and magnetic fields. That means that the four equations cannot be completely independent. We can combine these four equations into only two equations, by taking the value of the fields from the first two equations and inserting them into the last two equations. This is an example of one of the mathematical techniques that the physicist uses to investigate the consequences of a mathematical description of motion.

When Maxwell did just that he made an interesting discovery. He found that the magnetic and electric fields were both described by exactly the same equation. That doesn't mean that they are the same quantity, because they still have to satisfy all four of Maxwell's equations. But it does mean that whatever the electric and magnetic fields are in free space, where there is no matter present, they must both exhibit the same behavior in space and time. What made this discovery so remarkable was the particular form taken by the equation describing both fields. It was an equation with which Maxwell and all physicists are intimately familiar: the equation describing the motion of waves in space and time. The implication was clear: electric and magnetic fields could be transmitted through space as waves. Furthermore, the wave equation contains a term corresponding to the speed at which the wave travels, and in Maxwell's wave equation that speed took the form $\sqrt{\frac{k_e}{k_m}}$. It seemed that these two fundamental constants of nature determined all by themselves the speed of electric and magnetic waves through space. If we substitute into this expression our previous numerical values of $k_e = 8.988 \times 10^9$ and $k_m = 10^{-7}$, we find that the speed of propagation of electromagnetic waves is approximately 3×10^8 meters per second, the same as the speed of light! Maxwell had discovered the fundamental nature of light itself.

That light was a wave of some sort had long been assumed. Light could be diffracted by passing it through very tiny apertures, the way the path of any wave is bent whenever encountering obstacles comparable in size to the wavelength or the distance between two successive peaks of the wave. Light passing through a tiny pinhole spreads out beyond it just the way a water wave does in passing through an opening between two obstacles. A beam of light passing through a pair of small apertures, or slits, side by side produces a pattern of light and dark regions on a screen placed behind the openings, just as one would expect from the interference between two waves, one emanating from each opening. Light can also be bent or refracted in passing from one medium to another, the way acoustical waves and water waves are refracted in traveling through materials of varying density. White light can be broken up into all the colors of the rainbow by shining it through a prism of glass, suggesting that the various colors correspond to different wavelengths of light. In all of these phenomena, light behaves just like a wave, not only qualitatively but also quantitatively when compared with the behavior of other kinds of waves. It was also understood that light is associated with some form of energy. Intense sources of light on the earth emit copious quantities of heat in addition to light, and that heat is transmitted through objects and through empty space just as light is. The exact form of energy associated with light was not understood.

With Maxwell's discovery, the answers to these questions suddenly became clear. Light is an electromagnetic wave of alternating electric and magnetic fields perpendicular to one another and each perpendicular to the direction of propagation. The electric and magnetic fields oscillate sinusoidally in both space and time. Since the time variation in either field is what creates the other, they always occur together in the electromagnetic wave. They are phased so that the electric field reaches its maximum intensity when the magnetic field is zero, and vice versa, as the wave moves along. What we see as light consists of coupled electric and magnetic fields propagating freely through space, even space far removed from any matter and even space empty of all matter. An electric field represents a force exerted on electric charge and a magnetic field represents a force exerted on moving charge. When we say an electromagnetic wave propagating through space, what we actually mean is an electromagnetic force traveling through space. The force is there whether or not any charge is present for it to act on. When the electromagnetic wave encounters charge or current, it will exert a force proportional to the quantity of charge and the amount of current. The electric field can do work since it exerts a force in the direction of motion of the charge. The magnetic force is always perpendicular to the flow of charge and the magnetic field does no work. Yet both fields carry energy. The magnetic field fluctuates in time, creating an electric field that gets its energy from the energy stored in the magnetic field. The energy carried by the magnetic field is thereby transferred to the electric field, where it is available to do work.

Light then is all of these things: an electromagnetic wave propagating through space, coupled oscillating electric and magnetic fields that can exert an electromagnetic force, energy and momentum that can be imparted to charged matter. Light is only one such form of electromagnetic radiation. There are many others having wavelengths that differ from those of visible light. Taken altogether they comprise the electromagnetic spectrum, from very long wavelength radio waves, through microwave, infrared, visible and ultraviolet radiation, to very short wavelength x-rays and gamma rays. We live our entire lives awash in a sea of electromagnetic waves. They are everywhere and impinge on us constantly from all directions. We can detect their presence from the motions they impart to charges in materials. A radio or TV wave sets the electrons in an antenna or transmission cable into motion, producing currents that can be detected and amplified to let us "hear" and "see" the electric energy carried by the wave. Sunlight heats the earth by imparting kinetic energy to the electrons, atoms, and molecules in the materials it strikes. An antenna or detector pointed at the night sky picks up radiation across almost the complete electromagnetic spectrum. The whole universe in every direction we look is filled with electromagnetic waves, including a uniform background of electromagnetic waves that is usually explained as being a remnant from the primordial fireball that created the universe in the initial big bang.

What produces these propagating electromagnetic fields? Maxwell's equations give us the answer to that also. They are produced by accelerating charges. A charge at rest or moving uniformly does not radiate electromagnetic energy. But a charge that is accelerating, because it is changing either its speed or the direction of its motion, creates an electromagnetic field that propagates outward from the accelerating charge as a wave in space. Electric currents in a transmitting antenna generate an electromagnetic wave that travels through space to another antenna where it in turn induces currents exhibiting the same pattern as those in the transmitter, thereby relaying a radio or TV signal from one place to another. What causes the charges to accelerate in the first place of course is an electromagnetic field, or force, acting on them, but that electromagnetic field is itself produced by the motion of charge. Propagating electromagnetic waves owe their existence entirely to the acceleration of electric charge.

Here we have an interesting parallel between the material properties of mass and charge. In our discussion of the laws of motion, we saw that the mass of a body times its acceleration represents the force acting on it or the force that it can exert in acting on another body. Now we find that the acceleration of electric charge produces an electromagnetic force that is transmitted through space to where it can be exerted on other charged bodies. In each case, it is the acceleration of the material property—either mass or charge—that is caused by an applied force and that in turn creates a force that can act on other material bodies.

As in the case of mass, our description gives us little insight into the nature of charge. Like mass, charge is a property of matter defined in terms of the operations by which we measure its numerical value. Both mass and charge are merely numbers in a mathematical description of motion required to make that description reproduce what we actually observe. Nothing about the description can tell us where in the fundamental components of matter the elusive properties of mass and charge reside, or of what, more fundamentally, they might consist. As far as we know, they are not material things at all but only properties, only numerical labels that indicate how bodies move. In that regard they are qualities like color, which we now know is nothing more than the numerical value of the wavelength of light emitted by an object. When it comes to determining the nature of material properties, the physicist's world does not represent reality; it is simply a quantitative description of motion.

With the discovery of charge, however, the description takes on a whole new complexity. In addition to gravitational forces, we have to include the electromagnetic force. Our description has to account not just for the motion of material bodies due to their mass but due to their charge as well, and in addition it has to account for the motion of the electric and magnetic fields acting on the charges in matter. Electromagnetic energy has its own motion independent of matter, or rather on a par with matter. We are no longer dealing with instantaneous action at a distance. The electromagnetic force extends throughout the universe but it does not do so instantaneously. Any change in the configuration of charge at one point in space is not instantly felt as a force at every other point. The force has to be transmitted as an electromagnetic wave at the speed of light. Although light travels incredibly fast from a human perspective—almost three hundred million meters per second (about 186,000 miles per second)—it still takes about eight minutes for light to reach the earth from the sun and 100,000 years for it to cross our Milky Way galaxy just once, and millions and billions of years to span the distances between galaxies. The universe contains perhaps one hundred billion galaxies, and the time required for light to reach us from the farthest extent of the known universe is some 13 to 14 billion years. On a cosmic scale, light travels very slowly indeed and any electromagnetic disturbance is practically a localized event. Something happening in our own galaxy at the dawn of human consciousness might still be completely unknown to us. We are effectively isolated from what is happening in all but a tiny portion of the universe by the enormous time required for light to travel the vast distances involved.

There is another property of the electromagnetic force that also isolates us. Maxwell's equations tell us that for large distances the fields that make up an electromagnetic wave decrease in direct proportion to the distance from the source of the wave. The force that the fields can exert on charged matter diminishes the farther removed we are from the source of the wave. As the wave propagates outward in all directions, the energy it carries spreads out over space and the amount available at any one location falls off proportional to the square of the distance from the source. The electromagnetic fields produced by non-accelerating charges (charges at rest or steady currents) fall off proportional to the square of the distance and the energy in the fields falls off even more rapidly. It is this same diminution of the gravitational force with distance, coupled with its relative weakness, that isolates us from gravitational disturbances in other parts of the universe, even if we were to assume that gravity acts instantaneously everywhere.

Both gravity and electromagnetism are somehow transmitted through space, and nothing in our description suggests how that is accomplished. Maxwell was not satisfied to leave it at that. How could an electromagnetic wave—or any kind of wave—propagate through empty space? What was it a wave in, what medium was undulating or "waving"? We are accustomed to waves traveling in some kind of material medium, such as waves on a string that vibrates back and forth or sound waves in which the material making up the transmitting medium vibrates to and fro. But what exactly could be vibrating in empty space? Our description of the motion says that it is the electric and magnetic fields that are vibrating, that the fields themselves oscillate in space and time, perpendicular to the direction of motion, just like the string vibrating perpendicular to the wave traveling along it. The fields may seem like abstract mathematical concepts, but they are not. They represent forces that the space is somehow able to exert on charges and electric currents. If we doubt it, we can place charges and currents in the space and measure their acceleration as evidence that forces must be acting on them. If we suspend a charge anywhere in space, we do indeed find that it accelerates in response to the electromagnetic fields present almost everywhere. But to Maxwell and others immersed in the Newtonian tradition of space and time, space was not a physical thing; it was the empty framework—the void—through which everything else moved. How could space exert a force on anything? There had to be some medium present—some physical material, some kind of matter—to convey the force, through which the electromagnetic wave could be transmitted. So they borrowed the idea of the "aether" from the Greeks, who had filled the empty spaces of the universe with it in order to escape any void and to couple the motion of the celestial spheres.

It was an idea that made almost no physical sense, but one that held great promise philosophically in clearing up the mystery about absolute motion. To explain the relatively large value of the speed of light, the aether had to be extremely stiff or rigid and yet had to have almost no mass or density. The two requirements seem physically incompatible, even contradictory. In addition, the aether, in spite of its rigidity and inflexibility, had to allow material objects to move through it unimpeded. But, if the aether did exist, then perhaps it was the one thing at rest in the universe through which everything else moved. The aether could very well be the elusive reference frame for measuring absolute motion that Newton had sought in vain. All that remained was some way of determining motion through the aether, and light seemed to offer the perfect solution.

An observer moving through the aether in the same direction as light would measure the speed of light relative to himself to be the difference in their respective speeds through the aether, the same way that two cars traveling in the same direction move apart at a relative speed equal to the difference between their speeds along the highway. But an observer moving through the aether in the opposite direction as light would measure the speed of light relative to himself to be the sum of their respective speeds through the aether, in the same way that two cars traveling in opposite directions approach each other at a relative speed that is the sum of their speeds along the highway.

If everything in the universe moves through the aether, then the speed of light measured in different directions should have different values. By measuring the speed of light in all directions and taking the difference between the largest and smallest values, we should obtain our speed through the aether and hence ascertain our absolute motion through the universe. And that is just what physicists set out to do at the end of the nineteenth century. They would settle forever the question of what in the universe is moving and finally complete the mathematical description of motion begun by Newton. Instead they discovered an even deeper mystery, one that would eventually force physicists to revise all of their concepts about space and time and even material properties like mass and that would usher in the strange discoveries of the twentieth century.

Before stepping into the twentieth century, let's take one last look at the nineteenth century and how the attempt to understand the nature of heat ended up finding out something unexpected about the motion of matter and energy: for the very first time, motion was discovered to have an irreversible direction.

11

Heat and the Arrow of Time

The science that today we call thermodynamics was created in a relatively short span of time in the nineteenth century, partly as a result of efforts to learn how to design more efficient steam engines. The intent at the outset was to understand the puzzling nature of heat and the laws that govern its generation and use, but, as with all physical description, thermodynamics turns out to have a fundamental connection with motion. From the beginning, it has been one of the most practical of sciences, certainly one of the most practically motivated, yet from it have come two of the most fundamental principles of the universe, laws that govern the nature and use of energy and that transcend any assumptions about the detailed composition of matter.

No one has ever succeeded in deriving the laws of thermodynamics from more fundamental principles or from any other description of motion or assumptions about the nature of matter. The laws of thermodynamics appear to represent true first principles, more fundamental than any of the other descriptions of natural phenomena. The laws of thermodynamics describe the macroscopic world of our immediate senses and are entirely empirical. They serve as a constant reminder that the real world is the one outside our window, not the one we construct mathematically in our minds, however tidy and appealing that mental world may seem. Thermodynamics is the perfect model for how every quantitative physical theory must be constructed: *out of basic concepts defined operationally to express empirical observations.* We will examine it briefly here with that in mind. The concepts of thermodynamics do not rely on any assumptions about the nature and composition of matter, such as point charges or point masses or whether mass and charge are distributed continuously or as discrete units but are also not at odds with such assumptions. In fact, assumptions about the composition of matter and the motion of its constituents must be checked for consistency with the laws of thermodynamics, not the other way around.

We begin as we always do in physics, with certain intuitive concepts about the nature of the world growing out of our sense perceptions. The one that concerns us here is our impression of hot and cold derived from the sense of touch. We say that an object is hot or warm or cold based on how it feels to us when we touch it. Even if we do not make contact we may still sense how hot or cold an object is by how it feels to our skin when we are close to it, as when we sit next to an open fire or stand before a snowbank. Since antiquity it has been recognized that our sense of hot and cold is very imprecise and can be deceiving. Water that feels warm when we first touch it may quickly grow to seem less warm or even cool as we keep our hand in it. Water that otherwise might seem warm may strike us as either hot or very cool depending on whether we first immerse our hands in ice water or extremely hot water before touching it. Our sense of hot and cold is at best only relative and at worst can be completely misleading. Experiences like these helped reinforce the skepticism of the early Greek philosophers about the reliability of the senses.

Yet there is clearly something about the concept of hot and cold that is not entirely misleading and can even be made unambiguous. A metal rod placed in a mixture of ice and water and carefully measured is found to have a certain length. As the ice melts, the length of the rod is found not to change. When the ice has completely melted and the water is heated, the length of the rod continually increases until the water begins to boil. While the water is boiling, the length of the rod again remains constant. When all of the water has been turned into steam, the length of the rod once more increases as the steam is being heated. If the same rod is now reimmersed in a mixture of ice and water, we find that it returns to its original length and the entire process is repeatable with the same results.

In other words, there is a physical property of the metal rod, namely its length, that uniquely corresponds to the physical processes going on in the ice and water mixture. If we look more closely, we find that every dimension of the rod, not just its length but also its volume, behaves in the same fashion. Furthermore we find the same behavior exhibited by solids, liquids, and gases, with liquids and gases, in that order, generally showing a progressively greater change in volume than solids. The length of the rod correlates with our own sense of hot and cold: the greater the length, the hotter the material. This gives us a convenient operational way of assigning numerical values to what we mean by temperature.

We choose a physical system that can be easily reproduced and designate it as our thermometer. In the example above, it was a rod of a certain length made out of some specified metal. Most solids undergo relatively small changes in volume as they are heated, so as a practical matter we may want to choose instead a column of mercury or some other liquid in an evacuated glass tube as our thermometer. Or we could use a cylinder of gas free to expand against a movable piston. Or instead of measuring volume we might decide to measure the pressure exerted by a constant volume of gas. Whatever we choose as our thermometer, we designate one of its physical properties (length, volume, pressure, etc.) as the thermometric property that we will use to indicate temperature. We can establish a universal temperature scale by choosing easily reproducible physical systems as temperature standards and arbitrarily assigning a temperature value to each of them. A convenient standard is a mixture of ice and water at standard atmospheric pressure (the pressure exerted by the atmosphere at sea level), which is assigned a temperature of 32 degrees on the Fahrenheit scale and a temperature of 0 degrees on the Celsius scale. Another is the temperature of boiling water at standard atmospheric pressure, taken to be 212 degrees Fahrenheit and 100 degrees Celsius.

We then calibrate our thermometer by measuring the value of its thermometric property when it is in contact with each of the standard systems. The thermometric property of the thermometer changes when it is brought into contact with one of the standard temperature systems, and we have to wait until it has reached thermal equilibrium before measuring its value. Two systems are said to be in thermal equilibrium when the thermometric properties of each have quit changing after they are in intimate physical (thermal) contact.

We will illustrate the process of calibrating our thermometer for the Celsius temperature scale. Suppose the thermometric property is the length of a mercury column in a glass tube. Suppose further that L_0 is the measured length of the mercury column when the thermometer is immersed in an ice and water bath, and L_{100} is the corresponding value when it is immersed in boiling water. The length changes by an amount $L_{100} - L_0$ for a temperature change of 100 degrees on the Celsius scale. The change per degree is $\frac{L_{100}-L_0}{100}$. When we put our thermometer in contact with a substance whose unknown temperature, T, we wish to measure, we find that the length has a measured value L. The change in length from L_0 to L represents a change per degree of $\frac{L-L_0}{T}$. Equating these two expressions for the change in length per degree gives us

$$\frac{L - L_0}{T} = \frac{L_{100} - L_0}{100},$$

which we can rewrite in the more convenient form

$$T = 100\frac{L - L_0}{L_{100} - L_0}.$$

This equation expresses the temperature T measured by our thermometer whenever the length of the mercury column has the value L. The values of L_{100} and L_0 are fixed, once we calibrate our thermometer at the ice point and the boiling point of water. We can get the temperature of any other substance by simply measuring the value that the length comes to when the thermometer is placed in contact with that substance and allowed to come to thermal equilibrium. Note that if we put $L = L_0$ in the above equation we get $T = 0$, and if we put $L = L_{100}$ we obtain $T = 100$, just as we should. If we use a thermometer with some thermometric property other than length, then we substitute the measured values of that property in place of the corresponding lengths in our equation for temperature.

In this manner we can take our intuitive concept of hot and cold and turn it into a quantitative measure of what we mean by temperature. This is a classic example of an *operational definition*. We define temperature by the specific operations necessary to measure its quantitative value. Temperature then becomes what we measure with our thermometer. A detailed description of the thermometer, along with all of the operations by which we calibrate and use it, are an essential part of the definition. Temperature is just the numerical value we obtain when we apply our operational definition. Outside of that it has no meaning—unless we can show that some other interpretation is equivalent to the operational definition we have given.

Notice in particular that the zero point of the temperature scale is completely arbitrary. By our operational definition, we can have negative temperatures as well as positive ones. The zero point of temperature is different on the Celsius and Fahrenheit scales. On one it occurs at the ice point of water; on the other 32 Fahrenheit degrees below that. Negative values of temperature on the Celsius scale correspond to values of length less than L_0 such that $L - L_0$ is negative. Any temperature colder than the ice point of water is negative. Nothing about temperature defined this way tells us whether there is an absolute lowest (most negative) value of temperature or not. Later we will describe a way of establishing a temperature scale which does have an absolute zero value.

There are a host of practical issues that must be resolved in applying our operational definition of temperature. For instance, we have to compare thermometers that use different thermometric properties to verify that they all measure the same value for the temperature of a substance. We have to adopt certain thermometers as standards against which to compare other thermometers, in the same way that we compare other temperatures to the ice point and boiling point of water. If you look closely, you will see that we assumed a linear relationship between temperature and the thermometric property of our thermometer. In our example, any change in the length of the mercury column is assumed to be directly proportional to the change in temperature. This assumption of linearity has to be verified experimentally by comparing different types of thermometers, and ultimately it restricts the *temperature range* over which any one of them can be used. Clearly we would not use a mercury-in-glass thermometer anywhere near the melting point of glass or near the freezing or boiling points of mercury. For very high and very low temperatures, other thermometers and other thermometric properties are needed, and, to calibrate them at the temperatures for which they will be used, additional temperature standards are required.

All of these are among the many concerns addressed by the science and practice of thermometry, none of which need occupy us here. This is just a sketch of how we could take a qualitative notion of hot and cold and, by adopting an operational definition, turn it into a quantitative concept based on measuring numerical values of temperature. It is noteworthy that Lord Kelvin, one of the founders of thermodynamics, remarked that one never really understood a physical concept until it had been reduced to something that could be measured. The physicist's world is a mathematical description based ultimately, and only, on quantities that can be measured.

Equipped now with a way of measuring temperature, we are ready to construct an operational definition of heat. The question naturally arises: what exactly is heat? Before we can measure it, we have to have some idea what we are measuring. Intuitively, we think of it as being what makes a body hot. The presence of heat in a body raises its temperature; its absence lowers the temperature and leaves the body colder. Whether heat is a physical substance of some kind or a material property in addition to mass and charge or something else altogether, our intuitive concept cannot tell us for certain. All it suggests is that heat is a quantity that can be transferred from a hotter body to a colder one, changing the temperature of both. That concept is quite enough on which to base an operational definition.

We will simply define heat as the quantity (of whatever it is) required to change the temperature of a specified amount of some standard substance by one degree on our temperature scale, as measured by our thermometer. We will call the units of this quantity *calories*. For example, one calorie is the heat required to raise the temperature of one gram of water by one degree Celsius. Ten calories of heat will increase the temperature of ten grams of water by one degree Celsius or the temperature of one gram of water by ten degrees Celsius. To determine the number of calories of heat absorbed by any quantity of water, we multiply its mass in grams by its temperature change in degrees Celsius. We choose water as the standard substance in calorimetry (the measurement of heat) because it takes more heat to change the temperature of water than most other common substances. We say that water has a higher heat capacity, meaning the ability to absorb (or give up) more heat per unit mass for a given change in temperature, than many other substances.

Now that we know how to measure heat, we can use our operational definition to determine the quantity of heat required to change the temperature of any substance by comparing it with water. We do this most simply in practice by measuring the change in temperature of a substance as heat flows from it into a known quantity of water. The temperature change of the water tells us how many calories of heat were given off by the substance, from which we can find its heat capacity, the number of calories per gram per degree Celsius required to change its temperature.

Once we know that, we can find the heat entering or leaving the substance by just measuring the change in its temperature and using the measured value of its heat capacity. Note that heat is defined *only* in terms of *changes* in temperature. The definition says nothing about the absolute quantity of heat in a body or even whether such a concept makes any sense. It does not say that the quantity of heat in a body is proportional to its temperature but only that the quantity of heat *gained* or *lost* by a body is proportional to the change in its temperature.

When we apply the concepts of temperature and heat to our surroundings, we discover that materials differ greatly in their capacity to transmit heat. Some materials, notably metals such as silver and copper, are very good thermal conductors and transmit heat readily. Others, like wood and asbestos and Styrofoam, are thermal insulators and form a physical barrier to heat flow. By means of such thermal insulators, we can isolate a system thermally from its surroundings so that it does not gain or lose heat, a condition that we describe as *adiabatic*. Even though an adiabatic system will not absorb or give off heat, we can still change its temperature by performing physical work on it or by having it perform work on its surroundings. Recall from chapter 9 that to the physicist work means the amount of energy produced by a force acting through a distance. As an example, when we compress a gas in an insulated cylinder it gets hotter, as anyone who has ever used a bicycle pump knows. If we let the gas expand, it cools off. In the first case, we are doing work on the gas by exerting a force through a distance to compress it, such as might be accomplished by a movable piston in one end of the cylinder. In the other case, the gas does work on its surroundings as it expands by exerting a force on the piston to move it in the opposite direction.

We now have two ways of increasing the temperature of a substance—by having it absorb heat or by doing work on it—and two ways to decrease its temperature—by extracting heat from it or by having it do work on its surroundings. It begins to look as if there is some kind of equivalence between work and heat, and that turns out to be the case. Empirically we discover that there is a fixed ratio between the amount of heat required to raise the temperature of a substance and the quantity of adiabatic work (work done on a thermally insulated system) required to raise its temperature by the same amount and that this ratio is the same for all substances. One unit of heat is always equivalent to the same number of units of work, something we call the Joule equivalent of heat after the British physicist who first measured it. Heat can produce the same effect on a substance as the equivalent amount of work, from which we conclude that heat must be some form of energy.

Previously when we were discussing the laws of motion, we learned that the work done on a material body by an applied force could change its kinetic and potential energy, and that the total energy—kinetic plus potential—remained constant. What happens to the work done on the gas when we compress it adiabatically? It isn't transferred to the surroundings as heat, because the gas is thermally insulated from its surroundings. Yet it produces the same temperature increase in the gas as would an equivalent amount of heat. The work done in compressing the gas must be stored in the gas itself as some kind of internal energy. If we let the compressed gas expand adiabatically, we can recover this stored internal energy in the form of work done by the gas on its surroundings. As the gas expands, it exerts a force on the piston that can be used to impart motion, and hence energy, to its surroundings. So there is an equivalence between the adiabatic work done on a substance and the increase in its internal energy. As the internal energy of the substance increases so does its temperature, so temperature is also associated with internal energy.

Heat absorbed by a substance that is constrained so that it can do no work on its surroundings will also raise its temperature—the same as an equivalent amount of adiabatic work—and must also be stored by the substance as internal energy. Thus there are two ways of increasing the internal energy of a substance, just as there are two ways of increasing its temperature: by heating it or by doing adiabatic work on it. Now suppose both things happen simultaneously. Suppose the substance absorbs or gives off heat at the same time that we do work on it or it does work on its surroundings. Joule and others were able to establish a universal empirical relationship between changes in internal energy, heat, and work, known as the first law of thermodynamics. We can write this relationship quantitatively as

$$U = Q + W,$$

where U represents the *change* in the internal energy of a substance, Q is the quantity of heat *absorbed* or *given off* by the substance, and W is the work *done on* or *by* the substance. The way we have to read this equation is that heat flowing into a substance and work done on the substance by its surroundings are both positive quantities that produce an increase in its internal energy. Heat given off by a substance and work done by the substance on its surroundings are negative quantities that produce a decrease in its internal energy.

The first law of thermodynamics thus expresses the conservation of energy. It states that heat absorbed by a substance in the absence of any work ($W = 0$) is not lost but increases the internal energy of the substance. Likewise, in the absence of any heat flow ($Q = 0$), the work done on a substance is not lost but is stored in the substance as internal energy. The stored internal energy can be recovered in the form of heat and as work done by the substance on its surroundings. The interesting thing about the first law is that we have not had to say anything at all about the exact form taken by internal energy. Once we have defined what we mean by heat and work, the first law merely tells us that physical substances possess some kind of internal energy without saying anything more specific than that about what causes the energy or what form it takes. We learned at the outset that thermodynamics is a macroscopic description, completely empirical, one that assumes nothing about the microscopic composition of matter. We now know that the internal energy is due to the kinetic and potential energy of the constituents of matter, its mass and charge, moving under the influence of the electromagnetic force and the laws of motion, but the details of that description don't matter and are not part of the laws of thermodynamics. Yet the first law assures us there is some kind of motion, or the capacity for motion, going on inside of matter, reinforcing the view that motion is the defining feature of the material world. Not only is motion the chief external characteristic of matter, but it is a fundamental part of its internal makeup as well.

In addition, the first law places a fundamental constraint on the nature of all motion, internal and external. To see that, we can rewrite the first law in the modified form

$$U - Q - W = 0.$$

Each of the terms on the left represents a change in something. U is the change in the internal energy of a substance. The quantity $-Q$ represents the heat given off by a substance to its surroundings. The quantity $-W$ represents the work done by the substance on its surroundings, and hence the change in the mechanical energy—kinetic and potential—of its surroundings. In this form, the first law says that the sum of all the changes in the energy of a substance and its surroundings is zero. In other words, the total energy of a substance and its surroundings—and hence the whole universe—is constant.

This is the law of the conservation of energy. It comes out of the science of thermodynamics because, in order for energy to be conserved, heat must be included as a form of energy. In fact the first law illustrates the central role of heat in the interactions of material bodies. Lord Kelvin advanced the idea that the conservation of energy is a universal principle of nature, one that pertains to all forms of energy, in the same way that we had to say nothing about the form taken by the internal energy or the form taken by heat energy. This view gradually took hold after the middle of the nineteenth century. Today the conservation of energy is considered perhaps the most fundamental law of nature, the one that supersedes and underlies all the others. Later we will see that it is equivalent to saying that the total energy and mass of the universe is constant.

The first law of thermodynamics tells us that energy is neither created nor destroyed but is merely changed from one form to another. It places no constraints on the way we can transform or use energy, so long as the total amount of energy remains constant. Yet certain energy transformations involving heat are never found to occur in nature. As perhaps the most obvious example, we never observe heat flowing spontaneously from a colder body to a hotter one. In all natural processes, heat flows in the other direction, from regions of higher to lower temperature. There are no exceptions. We can extract heat from a substance and deliver the heat to its surroundings, cooling the one and heating the other—that is exactly what a refrigerator or an air conditioner does—but in order to accomplish that we have to do work and expend energy. No one has ever seen the water in a pond just spontaneously freeze on a warm day, heating the air above it or the ground below it. So long as the heat energy going to the air and the ground is the same as that leaving the water, such a process would be consistent with the first law of thermodynamics. The fact that it does not occur in nature suggests there is an additional principle at work, one known as the second law of thermodynamics, which expresses a fundamental restriction obeyed by all natural processes. Before stating the second law, however, we must first describe the absolute, or Kelvin, temperature scale. To do that, we want to consider a particular set of physical processes.

Consider an initial volume of some substance (such as a very dilute gas) in thermal equilibrium with a heat source at constant temperature. Since the substance and the heat source are at the same temperature, initially no heat flows between the two. Now let the substance absorb an amount of heat Q_1 from the heat source, by expanding its volume. Then remove the substance from the heat source and allow it to expand further, but this time adiabatically with no heat flow, cooling the substance until its temperature is reduced to that of a second heat source. Now consider the same two processes in the reverse order. First let the substance expand adiabatically, thermally insulated from its surroundings, cooling it until its temperature is reduced to that of the second heat source. Then with the substance in thermal contact with the second heat source let it absorb a quantity of heat Q_2 by expanding its volume until it has the same final volume as before. The substance has gone from the same initial volume and temperature to the same final volume and temperature by two separate processes. In the first it absorbed an amount of heat Q_1 at one temperature, and in the second it absorbed an amount of heat Q_2 at a lower temperature. Both processes have left the substance exactly the same—at the same final temperature and final volume. Whatever changes occur in the substance as a result of the one process must also occur as a result of the other. Since heat has been absorbed in both cases, there has to be some quantity involving heat that changes in exactly the same way for both processes.

To see what that quantity might be, let us examine the temperature scale proposed by Lord Kelvin, who suggested defining an absolute temperature for the two heat sources based on making temperature proportional to the heat absorbed from each. If we let T_1 and T_2 denote the absolute temperatures of the two heat sources, then Lord Kelvin suggested that the ratio of their absolute temperatures be the same as the ratio of the heat absorbed from each,

$$\frac{T_1}{T_2} = \frac{Q_1}{Q_2},$$

or in a slightly different form,

$$T_1 = \left(\frac{Q_1}{Q_2}\right)T_2.$$

In this way we can determine every other temperature in terms of some standard temperature, T_s,

$$T = \left(\frac{Q}{Q_s}\right)T_s,$$

where T_s is the arbitrarily designated temperature of some standard heat source and Q_s is the quantity of heat absorbed from it in the process described above.

It is customary to designate the ice point of water as having a temperature of 273.15 on the Kelvin temperature scale, so that the Kelvin temperature is given by

$$T = \left(\frac{Q}{Q_s}\right)273.15.$$

Since the Kelvin scale designates the ice point of water as 273.15 degrees and the Celsius scale as 0 degrees, then the Kelvin and Celsius temperature values differ by 273.15. A temperature of absolute zero on the Kelvin scale corresponds to –273.15 degrees Celsius.

This last expression tells us the meaning associated with a temperature of absolute zero. It is the temperature at which a substance would be unable to give off any heat to its surroundings ($Q = 0$). A substance at absolute zero can only absorb heat. No heat can be extracted from it, and absolute zero designates a unique direction to the flow of heat.

With absolute temperature defined in this manner, the ratio $\frac{Q}{T}$ becomes the quantity that has the same value for both of our heat transfer processes, namely,

$$\frac{Q_1}{T_1} = \frac{Q_2}{T_2},$$

where we have simply rewritten our original expression in this equivalent form. This ratio—the heat transferred divided by the absolute temperature at which the heat is exchanged—is called *entropy*, after the

Greek word for transformation. More properly, we should speak of a *change in entropy*, since the expression for entropy involves a change in the heat content of a substance. In the first process, the substance absorbed an amount of heat Q_1 at an absolute temperature of T_1, and the increase in the entropy of the substance was just $\frac{Q_1}{T_1}$. The entropy of the heat source decreased by an equal amount since it lost heat. In the second process, the entropy of the substance increased by $\frac{Q_2}{T_2}$ and the entropy of the heat source decreased by the same amount. We can find the change in entropy for every physical process. If the temperature is not constant, then we divide the complete process into increasingly many infinitesimal increments of temperature during each of which we can consider the temperature to be constant. We then find the quantity of heat transferred and the resulting change in entropy during each temperature increment and add all the results together using the methods of the integral calculus to find the total entropy change.

In the ideal processes described above, referred to as Carnot processes, after the French natural philosopher Sadi Carnot who first discussed them, the net change in entropy is always zero. The change in entropy of the heat sources is in each case just offset by the equal but opposite change in entropy of the substance. We refer to Carnot processes as ideal because we have ignored certain energy transformations and losses, notably those caused by friction. Although friction losses can be reduced, they can never be eliminated completely. As a result there are no true Carnot processes in nature, only approximations. The second law of thermodynamics states that for every natural process the total entropy of the universe must increase. The only processes that can occur in nature are those for which the combined entropy of all substances and their surroundings increases. We can decrease the entropy of a substance locally, but only at the expense of increasing the entropy by a greater amount somewhere else in the universe. The first law says that the only processes allowed in nature are those that leave the total energy of the universe unchanged. The second law says that of those processes that conserve energy, the only ones allowed are those that increase the total entropy of the universe. Just conserving energy is not enough. We must also increase entropy.

Now we can see why heat does not flow from a colder body to a hotter one. Suppose a quantity of heat left a body at one temperature and was absorbed by another body at a higher temperature. Energy has been conserved. The heat lost by the colder body is equal to the heat gained by the hotter one. But the change in the entropy of each body is the heat lost or gained divided by the corresponding absolute temperature. The colder body therefore has the greater change in entropy. The entropy of the colder body decreases by more than the entropy of the hotter body increases (the same quantity of heat divided by a lower temperature), leading to a net decrease in the entropy of the universe. Such a process would violate the second law. If the heat flow is in the other direction, the entropy of the colder body increases, since it now absorbs heat, and the increase is greater than the decrease in the entropy of the hotter body. In this case, the total entropy of the universe increases and heat flow is permitted.

The second law requires that heat flow from higher to lower temperatures. Heat, we say, flows downhill, tending always in the direction of lower temperatures. We can increase the temperature of a substance by heating it, but only by delivering some quantity of heat to a lower temperature somewhere else in the universe. By the first law, heat is a form of energy and represents the capacity to perform physical work on the surroundings. Heating a gas causes it to expand enabling it to push a piston and exert a force through a distance to cause motion. As the gas expands, the heat energy is converted into the kinetic and potential energy of material bodies. The higher the temperature, the greater the quantity of heat in the gas and the more work it can perform. As heat flows downhill, it loses its capacity to do work, and the increase in the entropy of the universe is a measure of that diminished capacity. At absolute zero, the heat energy in a substance is totally unavailable for work.

The second law of thermodynamics tells us that all processes in nature are irreversible. They can proceed only in the direction that causes entropy to increase. If a natural process could be reversed, the entropy of the universe could be decreased, in violation of the second law. The entropy principle provides a unique direction to time: the arrow of time, as Arthur Eddington called it. Time cannot suddenly go backward, or in the other direction, reversing all of the physical processes along with it. If that happened, the entropy of the universe would start decreasing toward the lower value it had in the past. Nature does not go backward in time, only forward. The value of entropy at any given instant of time is what divides the past from the present and the present from the future.

So far the second law of thermodynamics is the only physical principle we have encountered that designates a unique direction to time. If we reverse the direction of time in Newton's equation of motion or in Maxwell's wave equation, by replacing t everywhere by $-t$, we find that the equations themselves are unchanged. The very same equations describe the motion of material bodies and electromagnetic waves for time going in either direction. If we made a movie of a falling body or a beam of light, the mathematical description of the motion would be exactly the same whether we ran the movie forward or backward. We would have no way of knowing from the mathematical equations themselves which version of the movie represented the actual world.

If our description of motion is a correct one, there is no reason to suspect that the ball we dropped will not suddenly rise back up to our hand or that we will not suddenly grow younger and disappear altogether, to be replaced by our ancestors also living their lives backward. The second law of thermodynamics is the statement that such things are never observed to happen. They are simply inconsistent with everything that we know about the world. We conclude that Newton's laws of motion and Maxwell's equations are at best incomplete descriptions of motion. They admit processes that are absent from our world. The second law of thermodynamics is required to complete our description of motion. It takes its place alongside the conservation of energy as one of two fundamental principles of the natural world. Together they serve as the overarching constraints that must be satisfied by any more detailed description of motion.

12

Who's Really Moving and What's the Correct Time?

At the end of the nineteenth century, the only issue physicists had left to be resolved seemed to be the question of absolute motion—how to measure it and confirm the absolute nature of space and time. So much had been accomplished in the two centuries since Newton's *Principia*, and so much of that within recent memory, that many physicists at the time, basking in the triumph of Maxwell's equations, talked openly about the end of physics. There was just this one final loose end to tie up, and for that it was proposed to use Maxwell's equations themselves.

Following the discovery that light is an electromagnetic wave, it was presumed that light must propagate through the fixed framework of space, consisting of the aether, in which case measuring the speed of light would enable one to detect motion relative to the aether. The aether, then, becomes the stationary reference frame with respect to which all other motion can be measured, validating Newton's ideas about absolute space. An observer stationary with respect to the aether would measure the ordinary speed of light through the aether. We will designate the speed of light as c. A moving observer, traveling through the aether at some velocity v_0, would measure the speed of light to be $c - v_0$ in the direction of motion and $c + v_0$ in the opposite direction. In any other direction, a speed intermediate between these two values would be obtained. Motion through the aether would be confirmed by simply measuring the speed of light in all directions.

Even the speed and the direction of motion of the moving observer through the aether could be determined. The speed of light would be a minimum in the direction of motion and a maximum in the opposite direction. The difference between the maximum and minimum values would be twice the speed of the observer through the aether:

$$(c + v_0) - (c - v_0) = c + v_0 - c + v_0 = 2v_0.$$

It would be hard to conceive a more elegant or convincing demonstration of absolute motion, confirming once and for all the absolute nature of space and time. Newton would be vindicated, and that vindication would come not from mechanics and the laws of motion but from electromagnetism and Maxwell's equations, affecting, along with the laws of thermodynamics, a grand synthesis of all of physics.

Such a measurement had to await two things. The first was the discovery that light is an electromagnetic wave traveling through space. The second was the ability to measure very small changes in the speed of light. Light travels about 186,000 miles per second, or 3×10^8 (that is, 300,000,000) meters per second. The earth in its orbital speed around the sun is traveling a little less than 19 miles per second, or about 3×10^4 (that is, 30,000) meters per second. The earth's motion through the aether should be at least this large, but might not be any larger. To detect it would require measuring fractional changes in the speed of light of 0.0001, $\frac{3\times10^8}{3\times10^4}$, or 0.01 percent. Two Americans, Albert Michelson and Edward Morley, developed a technique capable of doing just that.

They shined a beam of light through equal distances perpendicular to one another onto two mirrors that reflected the beams back to the point of origin. Where the two reflected waves were in phase, that is, oscillating in the same direction at the same point, the light intensities added together to produce a bright region on the screen. Where the two waves were out of phase and oscillating in opposite directions, they canceled each other out and produced a dark region on the viewing screen. These light and dark regions constituted the *interference fringes*. When the apparatus was rotated through 360 degrees, any change in the speed of light in different directions would change the phase between the two reflected light waves and produce a shift in the location of the interference fringes on the screen. To isolate the apparatus from vibrations, which themselves could shake the instrument and cause the fringes to move around, they mounted the instrument on a massive block of stone floating in a pool of mercury. The experiment consisted of setting the stone block slowly rotating and walking around as it turned, peering through a telescope at the interference fringes to detect any shift in their position on the screen. The orbital motion of the earth alone should have been enough to produce easily observable shifts in the interference pattern.

Imagine the consternation when no such shifts were ever observed. The experiment was repeated at night, during the day, and at all seasons of the year. Surely at one of those times in one of those directions the earth had to be moving through the aether. The experiment was refined and repeated many times by Michelson and Morley and was checked by others, not once but again and again. It has become one of the classic experiments of physics, conducted many thousands of times, always with the same inescapable conclusion: no motion through the aether can be detected. This negative result constituted a monumental crisis for our understanding of motion, and for our understanding of the world—the first of many to come.

The initial reaction was to explain it away. Perhaps the earth was not moving through the aether after all was one conjecture. Perhaps the aether is being dragged along locally with the earth in its motion through space was another. That was not a very appealing idea, since it meant the aether was not truly stationary but could be moving at different speeds in different locations, but it might have explained the results. If it were true, we would expect to see a distortion in starlight when it travels through the interface or boundary between the moving aether around the earth and the stationary aether beyond. No such distortion is observed. It was even suggested that the length of the apparatus changed in the direction of motion as it was being rotated, in such a way that the change in length exactly compensated for the change in the speed of light, a suggestion that seemed bizarre at the time but later turned out to be prophetic. Other explanations were tried but all of them failed to account for the negative results.

Finally the Michelson-Morley experiment was accepted as establishing a totally unexpected and puzzling property of light: the speed of light is the same for all observers in uniform relative motion. A supposed stationary observer and one moving at constant velocity will both measure the same value for the speed of light in any direction. This means that light cannot be used to decide which observer might be stationary and which one moving. All that we can know is that they are moving at constant velocity (constant speed in a straight line) with respect to one another. Either one might be sitting still and the other in motion. Or they could both be moving but at different speeds. The dream of using light to detect absolute motion through Newtonian space suddenly vanished.

If the result of the Michelson-Morley experiment doesn't seem strange to you, it should. Imagine two observers, one inside a train moving at constant velocity along the tracks, the other one sitting on the ground beside the tracks. At the instant when the two observers are side by side, suppose each of them flashes a light toward the engineer at the front of the train. Which light flash will the engineer see first? According to the Michelson-Morley experiment, the speed of light is the same for all uniformly moving observers, including the engineer, and he should see both light flashes at the same time. But surely that cannot be. For the observer on the train, the light only has to travel the length of the train between herself and the engineer. For the observer beside the tracks, the light must travel that same length plus the additional distance that the train has moved forward by the time the light arrives at the front of the train. How could the light flashes possibly arrive at the same time? Yet according to the Michelson-Morley experiment they should—and do. Since the two observers disagree about the distance traveled by the two light flashes, yet agree about the speed at which they travel, then they must also disagree about how long the light has to travel in each case. Clearly there must be something very wrong with our commonsense notions about space and time, and that is what we want to examine next.

We first have to be very clear about what our commonsense ideas of space and time actually are. Since physics is a description of motion, the question we must answer is how two observers in uniform motion will compare their measurements of such things as distance, time, velocity, and acceleration. To answer this question, we will engage for a while in the kind of analysis that the mathematical physicist employs all the time. We will use it as an opportunity to construct a mathematical description and then use that description to help us think about the question of how moving observers compare measurements of space and time.

We will call our two observers O and O' for simplicity. For the sake of illustration, we will assume that O is at rest and O' is moving horizontally at constant velocity v_0 relative to O. All we can really know of course is that the two observers are moving apart. We could just as easily assume that O' is at rest and that O moves at a speed v_0 in the opposite direction. The only thing that matters here—because it is the only thing we can know—is the relative motion between the two. When we say that O is at rest, we just mean that we will ignore any motion of O except for the relative motion between O and O'. Each observer carries a meter stick for measuring distance from his or her position to any other location. The relative motion between the two observers could be in any direction; but for the sake of illustration we will take the motion to be in the horizontal direction.

Now suppose that O' uses her meter stick to measure the fixed distance from herself out to some object traveling along with her located at point P in the direction of motion, and suppose she obtains the value x'. (Here x' merely represents whatever number she gets for the value of her measurement.) When O measures the distance from himself out to the same point, he obtains the value x. How are these two distance measurements related? Clearly they differ by the distance d between the two observers, or

$$x = x' + d,$$

which just states symbolically that the distance x from O to P is the same as the distance d from O to O' plus the distance x' from O' to P. This equation expresses mathematically what our commonsense understanding of straight line distance tells us, namely that, if we break a total distance into two parts, the total is just the sum of the parts. Nothing could be more apparent or straightforward.

Since O' is moving away from O, the distance between them increases with time. We can rely once again on our commonsense understanding of the world to tell us exactly how d increases with time. If O and O' are moving apart horizontally at a constant speed v_0, then the total distance between them after an interval of time t will be the product of the speed and the elapsed time, or

$$d = v_0 t.$$

This is the same expression we have already made use of several times in our discussion of motion. In this expression $d = 0$ when $t = 0$, which signifies that we start measuring time when O and O' are abreast or side by side—we start our clock or stopwatch just as O' passes O. Putting this last expression for d into the previous equation gives us

$$x = x' + v_0 t.$$

This equation describes how observers O and O' would compare the results of their respective measurements of the horizontal distance from themselves to an object located at point P and traveling along with O'. It tells us how to transform the measurements made by either one of the observers into those made by the other and is referred to as the *Galilean transformation*, after Galileo who was the first to describe the uniform motion that it represents. It embodies nothing more than our common-sense understanding of distance and velocity. Remarkably, as we shall see, it is *simply wrong*.

Next we want to find out how measurements of velocity made by each of these two observers would compare or transform. Up to now we have considered the object at P to be moving along with O' so that the distance from O' to P was fixed. Now let's assume that the object can move freely in the horizontal (x and x') direction and change its position with respect to both O and O'. (We will limit all motion to the horizontal direction for simplicity.) Then O' can determine the speed of the object by measuring the change in its position (the change in x') in a given interval of time. Observer O can also measure the change in the position of the object (the change in x) in the same interval of time and determine the speed of the object from his measurement. How will the two measurements compare? We can use our expression for the Galilean transformation to find out.

Suppose O measures the change in position to be δx, where as before we employ the Greek letter δ to indicate the change in a given quantity. Then

$$\delta x = \delta(x' + v_0 t),$$

where we have replaced x by its equivalent value from the Galilean transformation, as we did in the equation above. We can calculate the change in the right side of this expression by using the fact that the change in the sum of two quantities is the sum of the change in each quantity, or

$$\delta x = \delta x' + \delta(v_0 t).$$

Since the velocity v_0 is constant and does not change, the change in the product $v_0 t$ is just v_0 times the change in t, or

$$\delta x = \delta x' + v_0 \, \delta t.$$

Dividing both sides of this equation by δt gives

$$\frac{\delta x}{\delta t} = \frac{\delta x'}{\delta t} + v_0.$$

The ratio $\frac{\delta x}{\delta t}$ is just the change in the position of the moving object per unit time, or its speed v, as measured by O; and $\frac{\delta x'}{\delta t}$ is the corresponding speed v' of the object measured by O'. We can rewrite the last equation as

$$v = v' + v_0.$$

This last expression tells us how the two separate measurements of velocity, one by O and the other by O', will compare or transform between the two observers.

A little thought shows that this equation also obeys our basic notions of the world. If O' throws a ball forward at 30 miles per hour while moving away from O at 60 miles per hour, the ball will move away from O at 90 miles per hour, the sum of the two speeds, just as we would expect. What if O' throws the ball backward, that is, toward O, at 30 miles per hour while moving away from O at 60 miles per hour? Then the ball should move away from O at 30 miles per hour, the difference between the two speeds. Does our velocity transformation equation also handle that case? Yes, it does, because, if the ball is thrown backward, then the change in its position as measured by O' will be negative (from larger values of x' to smaller ones), and the velocity v will be negative, which means that we will be taking the difference between the two speeds just as our senses would dictate. What about the situation where O and O' are moving toward one another instead of apart? Is that case also covered? Again, it is, since in that case by the same argument the velocity v_0 is negative instead of positive (O' travels from larger values of x to smaller values).

This velocity transformation equation expresses what came to be called the Galilean principle of relativity. It so thoroughly agrees with our common sense about how the world works that for almost three centuries it was the cornerstone of mathematical physics, and no one ever dreamed of seriously questioning it. What a shock then to discover that, when we try to use it to describe the motion of light, we end up with a conclusion that flies in the face of observation.

Suppose that instead of measuring the speed of a material object, two observers in uniform relative motion decide to measure the speed of a beam of light. Instead of throwing a ball forward, observer O' shines a beam of light forward and measures its speed. According to Galilean relativity, observer O' should obtain a different value when he measures the speed of the same light beam. The two measurements should differ by v_0, the speed of the relative motion between the two observers. If we assume that O is at rest in the aether and measures the speed of light to be c, then by the Galilean velocity transformation O will measure the speed of the same beam of light to be

$$v' = c - v_0.$$

It was this difference in the two speeds that the Michelson-Morley experiment was designed to look for in order to determine the absolute motion of an observer through the aether.

But no difference was ever found, no matter what our common sense tells us to expect. In a sense Newton was right; there is something absolute about motion, but it is not to be found in the absolute nature of space and time. It is just this: *the speed of light in a vacuum is the same for all observers in uniform motion.* It is the speed of light that is absolute. This principle, which has come to be called simply *the principle of relativity*, is apparently the cornerstone on which the description of all motion must be based. Einstein made this principle the starting point for his theory of special relativity, which he published in 1905. The use of the term *special relativity* denotes that it deals with the special case of uniform relative motion. The more general case of two observers in nonuniform or accelerated relative motion is dealt with in the theory of general relativity.

Where could we have gone wrong in our analysis? At what point did our thinking lead us astray? In comparing the values of velocity measured by two moving observers, we assumed that there would be no difference in the values they measured for elapsed time. After all they both use clocks, and we give them identical clocks. The only difference between the two clocks is that one is moving with respect to the other. It seems reasonable to expect a moving clock to measure time the same as one at rest, doesn't it? Isn't time universal and absolute? Strangely enough it turns out the two clocks don't measure time the same. Moving clocks run more slowly, as we will soon discover.

We can begin to see why by asking how we would go about examining whether two clocks in relative motion actually do run the same. Suppose observer O wants to investigate whether the elapsed time measured by his (stationary) clock agrees with the elapsed time shown on a clock moving with observer O' at velocity v_0. He could read his clock at two different times and attempt to read the moving clock at the same times to see if they agree. Suppose he looks at both clocks at the same instant. The moving clock is farther away, and it takes some amount of time for light to travel from it to observer O. When he looks at the two clocks at the same instant, the moving clock will show an earlier time than the stationary clock. It will show the earlier time that the clock was reading when the light began its journey to his eyes. At some later instant he again looks at both clocks. Now the moving clock is even farther away and will show a time that is earlier by an even greater amount. Thus the two clocks won't measure the same interval of time. The moving clock will show a shorter interval of elapsed time and appear to be running slower. It seems that we have some more work to do in order to determine how moving observers will compare their measurements of time.

The work has already been done for us, by Albert Einstein and legions of successors. Einstein replaced the Galilean transformation with a modified scheme known as the Lorentz transformation, for Dutch physicist Hendrik Lorentz, who used it in the late nineteenth century but for entirely different purposes. The Lorentz transformation assumes that moving observers will obtain different values not only for the measurement of distance but also for the measurement of time. Einstein proposed the new rules for the same reason that any physicist does—because they agree with what we observe, in this case the observation expressed by the principle of special relativity. Whether they seem to make immediate sense or not is beside the point. The only "sense" they have to make is to agree with our observations of the world. Our common sense is derived from our limited experience of the world, and none of us has ever experienced motion at the speed of light. The final criterion of any mathematical scheme in physics is not whether it agrees with our previous notions of the world but how well it describes the observed motion.

We won't derive the equations for the Lorentz transformation here, though they can be easily derived from the principle of relativity stated earlier. Instead we will just note that they are

$$x = \varepsilon\,(x' + v_0 t)$$

and

$$t = \varepsilon\left(t' + \frac{v_0 x'}{c^2}\right),$$

where ε is a quantity that depends on the ratio of the relative velocity v_0 to the speed of light c, namely

$$\varepsilon = \frac{1}{\sqrt{1 - \frac{v_0^2}{c^2}}}.$$

or 1 divided by the square root of the quantity $1 - \frac{v_0^2}{c^2}$.

Here x' and t' are the values of distance and time measured by observer O' using her meter stick and clock, both of which are moving at a constant speed v_0 with respect to observer O. The corresponding values of distance and time measured by observer O with his meter stick and clock are indicated by x and t. The Lorentz transformation states that the two observers measure different values for both distance and time.

The first thing to notice is that, if $v_0 = 0$ (no motion between the two observers), then $\varepsilon = 1$ and the Lorentz transformation reduces to $x = x'$ and $t = t'$. In other words, both observers obtain the same values for the measurement of distance and time, just as they should in that case. Whenever v_0 is not zero but is very small compared with the speed of light—which is the case in all our direct experience of the world—then ε is approximately equal to 1 and the Lorentz transformation becomes the same as the Galilean transformation,

$$x = x' + v_0 t'$$

and

$$t = t'.$$

We have developed our Galilean perceptions of distance and time for this very reason—precisely because we are limited to speeds that are small compared with the speed of light. Even for an object moving as fast as 1 percent the speed of light, the value of ε is 1.0005 and this is close enough to 1 that the Galilean transformation is still a very good approximation. And that speed is already one hundred times the speed of the earth in its orbit around the sun. Finally, note that ε is always greater than 1 when v_0 is not 0. As v_0 approaches the speed of light, then $\frac{v_0}{c}$ approaches 1 and ε as a result becomes large without limit. At speeds comparable to the speed of light, the Galilean transformation clearly would not work.

Instead of just blindly accepting the Lorentz transformation, we want to test it to see if it really works to explain the principle of relativity. We want to check whether it correctly predicts that two uniformly moving observers will indeed measure the same value for the speed of light. As we did before, we will calculate the change in position and the corresponding change in time measured by both observers, this time using the Lorentz transformation equations. For δx and δt we get

$$\delta x = \varepsilon\,(\delta x' + v_0 \delta t')$$

and

$$\delta t = \varepsilon\left(\delta t' + \frac{v_0 \delta x'}{c^2}\right),$$

where we have made use of the fact that ε, v_0, and c are all constant values that don't change during the measurement. Dividing the first equation by the second one gives

$$\frac{\delta x}{\delta t} = \frac{\delta x' + v_0 \delta t'}{\delta t' + \frac{v_0 \delta x'}{c^2}}.$$

Dividing both the numerator and denominator of the right side by $\delta t'$ gives

$$\frac{\delta x}{\delta t} = \frac{\frac{\delta x'}{\delta t'} + v_0}{1 + \frac{v_0}{c^2}\frac{\delta x'}{\delta t'}}.$$

Recognizing that $\frac{\delta x}{\delta t}$ is just the velocity v measured by observer O, and $\frac{\delta x'}{\delta t'}$ is the velocity v' measured by observer O', gives us

$$v = \frac{v' + v_0}{1 + \dfrac{v_0 v'}{c^2}}.$$

This expression tells us how velocities transform between the two moving observers. The numerator on the right side of this equation is the same result that we obtained using the Galilean transformation previously, but now we have the additional quantity in the denominator. If v_0 and v' are small compared with c, then the denominator is close to 1 and the Galilean transformation in that case is again a good approximation.

Now suppose that O' measures the speed of light and obtains the value $v' = c$. What value will observer O obtain from his measurement? We can find out by substituting c for v' in the last equation to obtain

$$v = \frac{c + v_0}{1 + \dfrac{v_0}{c}}$$

From this equation, we can determine that $v = c$ using the following steps,

$$v = \frac{c + v_0}{1 + \dfrac{v_0}{c}} = \frac{c + v_0}{\dfrac{c}{c} + \dfrac{v_0}{c}} = \frac{c + v_0}{\dfrac{c + v_0}{c}} = (c + v_0)\left(\frac{c}{c + v_0}\right) = c.$$

So both observers obtain the same value for the speed of light as required. If we substitute the value $v = c$ into the left side of the velocity transformation equation, we also obtain $v' = c$.

Unlike the Galilean transformation, the Lorentz transformation does correctly predict that all uniformly moving observers will measure the same value for the speed of light; but it does so at the expense of having them *disagree* about their measurement of both distance and time. Distance and time are no longer absolute quantities as Newton assumed but have values that depend on the relative motion of the observer.

Let us examine how two observers in uniform motion will compare their respective measurements of time. Suppose both O and O' have identical clocks, and we place the moving clock at the position of observer

O', that is, at $x' = 0$ (she is carrying it in her hands, for instance). With $x' = 0$, the second of the Lorentz transformation equations simplifies to

$$t = \varepsilon t',$$

which gives the relationship between the values of time measured by the two clocks. Let t_1 and t_2 be two consecutive readings of his clock made by observer O, and let t_1' and t_2' be the corresponding readings made by O' using her clock. Then according to our last expression

$$t_1 = \varepsilon t_1'$$

and

$$t_2 = \varepsilon t_2'$$

so that, subtracting these two equations,

$$t_2 - t_1 = \varepsilon (t_2' - t_1').$$

But $t_2 - t_1$ is the elapsed time measured by O, and $t_2' - t_1'$ is the corresponding time interval measured by O'. Since ε is greater than 1, this last equation says that the time between two events measured by the stationary clock will always exceed that measured by the moving clock, so that the moving clock will appear to be running slower. Observer O will conclude that his clock is running faster than the clock moving with O'. Observer O' will conclude just the opposite: that her clock (which is stationary with respect to her) is running faster than the clock that is moving uniformly with respect to her. In either case, it is always the moving clock that appears to be running slower to either observer.

Is it really running slower, or does it just appear to be running slower? If we adopt the point of view that all we can know about the world is what we can actually measure, then the moving clock is *really* running slower, since it always measures a time interval shorter than the interval measured by the stationary clock. This was the point of view taken by the mathematical physicist in the twentieth century and represented a fundamental shift from the Newtonian concept of absolute values for space and time. The reality underlying the physical world has to be determined by measurements. The Lorentz transformation tells us how measurements of position and time made by uniformly moving observers must compare for them to measure the same value for the speed of light.

Each observer will measure the other's (moving) clock to be running slower, so it really is running slower. The older, incorrect Newtonian view was not based on measurement. Newton merely assumed what he considered to be obvious without analyzing how one would have to go about comparing measurements of both space and time made by moving observers.

In the same fashion, we can show that meter sticks are shorter when they are moving; distance shrinks in the direction of motion. Suppose that observer O' measures the position of a meter stick accompanying observer O at some time t' on her clock, and suppose she obtains the values x_2' and x_1' for the location of the ends of his meter stick. (Note that this meter stick is in motion with respect to O' since she is moving with respect to O.) Observer O makes the same measurements and obtains the values x_2 and x_1 for the location of the ends of the meter stick. From the first of the Lorentz transformation equations, we have that

$$x_2 = \varepsilon\,(x_2' + v_0 t')$$

and

$$x_1 = \varepsilon\,(x_1' + v_0 t').$$

By subtracting these two equations, we obtain the result

$$x_2 - x_1 = \varepsilon\,(x_2' - x_1').$$

Now $x_2' - x_1'$ is the length of the moving meter stick measured by O', and $x_2 - x_1$ is the length of the stationary meter stick measured by O. Since ε is always greater than 1, the length measured for the stationary meter stick is greater than that measured for the moving meter stick. Each observer will conclude that the other's (moving) meter stick is shorter than his or her own. Again we have no choice but to conclude, as physicists, that this contraction is real. Whenever we measure the length of a moving meter stick, we find it to be shorter than the same meter stick when it is not moving. And whatever we measure represents the only reality we can ever know or experience.

The contraction of length and dilatation of time account for the surprising result that all uniformly moving observers measure the same value for the speed of light. Measurements of distance and time are "adjusted" by motion. Lengths contract and time slows in just the right way to make the measured distance traveled by a beam of light in a given interval of time always the same for all observers in uniform motion. This effect went unnoticed for so long because it depends on the value of ε which in turn depends on the ratio $\frac{v_0}{c}$. Obtaining a contraction or slowing down of only 1 percent requires motion at greater than one tenth the speed of light, a velocity unattained by macroscopic material bodies in nature and unattainable by any means in the laboratory until the discoveries of atomic and nuclear physics in the twentieth century. We don't normally notice the contraction of distances or the slowing down of time because we move so slowly. But when we measure the speed of light we do observe them, and the ability to make more and more exact comparisons of how fast light travels—as in the Michelson-Morley experiment—is what led to the realization that our previous ideas of space and time as absolute and unchanging quantities were incorrect and would have to be abandoned. The new concepts of space and time in turn introduce changes in our previous description of motion.

As we might expect from the fact that they describe the motion of light, Maxwell's equations are unaffected by the Lorentz transformation. If we replace distance and time measured by one observer with those given by the Lorentz transformation for a uniformly moving observer, Maxwell's equations take on exactly the same mathematical form for both observers. The equations for observer O' are the same as those for observer O, with x and t replaced by x' and t'. The same is not true however for Newton's equation of motion. There, if we replace x and t with the corresponding values of x' and t' given by the Lorentz transformation, we find that the equation of motion for O' is not the same as for observer O. The description of motion differs for the two observers.

To make Newton's equation of motion invariant to (unchanged by) the Lorentz transformation, the mass of a moving body cannot be constant but must depend on its relative velocity according to the relationship

$$m = \varepsilon m_0 = \frac{m_0}{\sqrt{1 - \frac{v_0^2}{c^2}}}.$$

In this equation, m_0 is the mass of the body measured when it is not moving relative to the observer, the so-called rest mass, and m is its mass when it is moving at a speed v_0 relative to the observer. The mass of a body increases as its speed increases, gradually at lower speeds but rapidly near the speed of light. As a result, progressively greater force is required to accelerate it to higher speeds or to slow it down. Normally we encounter bodies moving too slowly for this effect to be measurable. In addition, Einstein also found that, for energy to be conserved for both observers, the total energy E of a material body had to be related to its mass according to the equation

$$E = mc^2 = \varepsilon m_0 c^2 = \frac{m_0 c^2}{\sqrt{1 - \frac{v_0^2}{c^2}}}.$$

This result is particularly significant; it says that there is a fundamental equivalence between mass and energy. Even at rest ($v_0 = 0$) a body possesses a total energy of $E = m_0 c^2$, called its rest mass energy. We have to modify the conservation of energy to include the energy associated with the mass of the universe. Assuming that the total energy of the universe is constant amounts to assuming that its equivalent rest mass is constant.

All of these results—the Lorentz transformation, the contraction of length and the slowing down of time, the increase of mass with increasing speed, and the equivalence of mass and energy—are direct consequences of the special nature of light. Putting it another way, they are all consequences of matter having the dual properties of mass and charge. The property of charge gives rise to electromagnetism, and the property of mass governs the acceleration of matter under the influence of forces. The two are joined through the concept of energy. Light is electromagnetic energy generated by the acceleration of charged matter, and matter possesses energy by virtue of its mass. But, unlike matter, light exhibits the unique property that it travels at the same speed in a vacuum for all observers in uniform relative motion.

These results lead us to an important conclusion, namely that the speed of light represents an upper limit to the velocity of any material body in the universe. As an object approaches the speed of light, the value of the ratio $\frac{v_0}{c}$ approaches 1, and the value of ε increases without limit. Consequently the mass of the object becomes larger and larger without limit, which is what we mean by the term infinite, as do the force and energy required to accelerate it, making it impossible for matter to ever attain the speed of light. Even if it could, time measured by a clock moving relative to an observer at the speed of light stops, which means that all physical processes associated with matter cease at that speed; and all lengths in the direction of motion shrink to zero so that anything in the material world traveling at the speed of light would be unobservable. At the speed of light, the material world would simply come to a stop and disappear. Matter can only *approach* the speed of light but can never quite reach it, and hence the speed of light in a vacuum constitutes an upper limit to how fast we can transmit mass and energy from one place to another.

How, then, you may be asking, can light travel at the speed of light? What makes light special in this regard is that it represents energy without any material substance. Light can travel at the velocity c because it consists of no matter whose dimensions can contract and shrink to zero. It also possesses no mass that must be accelerated by the application of an infinite force or the expenditure of infinite energy. A beam of light always moves at its characteristic speed. It doesn't accelerate or decelerate. There is no necessity to get it up to speed since it starts off moving at constant velocity. Hence it does not behave like a material substance at all.

Our concepts of simultaneity and causality also have to be modified to bring them into agreement with the unique property of light. Moving observers will not necessarily agree on whether two events occur simultaneously, which calls into question our ability to associate events in a causal relationship. Imagine a train traveling along straight tracks at a constant speed with an observer O' seated at the exact center of the train. Imagine also an observer O seated beside the tracks as the train is passing by. Now suppose that, at the instant the two observers are directly opposite one another, lightning strikes the ground at each end of the train. Both observers will see the light propagating from the ends of the train to their eyes at the same speed, since light travels at the same speed for all uniformly moving observers. The two flashes of light travel the same distance to reach observer O. He will see them at the same time and will conclude that the two flashes occurred simultaneously. Not so for observer O'. She is traveling toward the flash of lightning that occurred at the front of the train and away from the flash of lightning coming from the rear of the train. Both flashes of light travel at the same speed, and hence the light from the front of the train reaches her first and she concludes that the two events were not simultaneous but were displaced in time. If the train had been traveling in the opposite direction, she would have concluded that the two events occurred in the reverse order. The observed order of the events will depend on the direction of her motion.

This simple straightforward example raises the troubling question of whether the concept of simultaneity has any meaning. At the very least simultaneity is not absolute, just as space and time are no longer absolute in the new relativistic physics but must be judged according to the motion of a given observer relative to that of other observers. Two events that occur simultaneously to observer O will in general not be simultaneous to any other observer who is moving uniformly with respect to O. Moving observers may also disagree about the order in which two nonsimultaneous events occur. When they do, both may be deemed correct, since each judges the order of occurrence by the order in which the events are observed to occur. Sticking with the point of view that all we are allowed to know about the world is what we can observe (measure), both observers are correctly recording their observations (measurements). There is no absolute way to choose between them. The concept of causality becomes limited by our inability to associate an absolute order to the occurrence of two events. In the case of events A and B that are always associated, whether we say A causes B or B causes A may very well depend on the relative motion of the observer. With no way of choosing which, if either, of two observers is at rest and which is really moving we have to abandon any absolute meaning associated with the concept of causality.

By now the theory of special relativity has stood the test of time. In every case where it is possible to check it by actual measurement, it has been found to provide the correct description of motion. It is a way of looking at the physical world that makes sense out of our observations. The Lorentz transformation accounts for the result that the speed of light is the same for all observers in uniform relative motion and explains the outcome of the Michelson-Morley experiment. The contraction of length and the slowing down of time explain the finding that radioactive matter decays more slowly the greater its speed with respect to the observer, in precise agreement with the results derived from the Lorentz transformation. The mass of an electron accelerated to higher and higher speeds increases according to the equation $m = \varepsilon m_0$. The energy required to accelerate a beam of electrons goes up accordingly, and the electric bill of a modern particle accelerator provides economic testimony to the correctness of relativity physics. Mass is converted to energy in the fission and fusion of atomic nuclei, electromagnetic energy can be converted to mass in the production of particles of matter, and mass converted into electromagnetic energy in the annihilation of particles and anti-particles, all in agreement with the famous $E = mc^2$ equation. Each of these predictions of the theory of relativity has been observed and verified countless times. None of its predictions is contradicted by observation. In every case where it could be falsified, the consequences and predictions of the theory have proven to be correct. By now it is an inescapable conclusion that the concepts of space and time represented by the Lorentz transformation constitute a more correct description of uniform relative motion. We can never go back to our older ideas of Galilean relativity. We have acquired a new common sense, one informed by the success of our mathematical description, and, although it takes some getting used to, it is now an indispensable part of our understanding of how the world works.

What comes out of all this is that the laws of nature take the same mathematical form for all uniformly moving observers. We say that all uniformly moving reference frames—all inertial reference frames—are equivalent for describing the laws of nature. That is the fundamental principle on which Einstein based the theory of special relativity and serves as an equivalent statement of the principle of relativity. There is no way from the observation of any physical phenomena—including the speed of light—to distinguish whether either of two uniformly moving observers is at rest. All that we can know is that they are moving relative to one another. The laws of nature consist of the mathematical description of motion, and all observers in uniform motion will arrive at the same mathematical form for that description. If not, we would be able to distinguish between observers on the basis of those differences and to use that distinction to claim a difference in their absolute motion. We might choose, for example, to declare the observer for which the description of motion was the simplest to be the one at rest.

We can make our description of motion—Maxwell's equations and Newton's equation of motion—assume the same mathematical form in all uniformly moving reference frames but only by accepting the Lorentz transformation as the correct comparison of independent measurements of space and time made by two observers. And that means accepting that meter sticks contract in the direction of motion and moving clocks run slower, that the mass of a body is not constant but increases with increasing speed, and that there is an equivalence between mass and energy. It means accepting that all we can know about quantities like space and time is what we can actually measure and giving up the concept of a separate absolute and universal reality. The real distance between two points is what we measure it to be, and if we *measure* it to be shorter when it is moving then it *is* shorter. The actual time between two events is that measured by a clock, and if we measure it to be shorter when the clock is moving then it is shorter. It also means accepting that simultaneity and causality are no longer absolute but depend on the relative motion of the observer.

There is an irony in calling it the theory of relativity. It is often regarded popularly as having made everything relative to the observer. Moving observers will disagree on their measurements of distance and time, will measure different speeds for a uniformly moving object, will also measure different values for its mass and energy, and will even disagree about the simultaneity of two events or about which one caused the other. Nothing is absolute in our observation of the world; everything is relative to the state of motion of the observer. But that misses the real point. The theory of relativity singles out the one thing about the natural world that is absolute: *the description of motion takes the same mathematical form for all observers in uniform motion*. The laws of nature are not relative but are absolute and universal. The constancy of the speed of light and its role as the upper limit to all motion is one of those laws.

Nothing about uniform motion can be used to infer absolute motion. The hope of using light to detect motion through space loses out to a more fundamental principle. We should not be surprised by this outcome. Even Newton realized that the laws of motion could not be used to distinguish between being at rest and moving at constant velocity. It should not surprise us that the laws governing the motion of electromagnetic waves are likewise unaffected by uniform motion through space.

But what about nonuniform motion? Newton argued that the presence of forces could be used to indicate the absolute acceleration of a body in space, making a distinction between uniform and nonuniform motion. There is no difference between a body at rest or one moving uniformly because there is no force acting on either. But when the body undergoes nonuniform motion there must be a force acting on it, and the presence of that force establishes that it must be accelerating with respect to some absolute point of reference. Yet we have determined that an electromagnetic wave, like light, which is nothing more than an electromagnetic force propagating through space, cannot be used to detect absolute motion. As a result we need to examine Newton's argument more closely.

13

Curved Space and the New Gravity

The results of special relativity tell us that the laws of physics are the same for all observers in uniform motion. Since every observer can carry along meter sticks and clocks for measuring distance and time, then each observer represents a particular reference frame with respect to which all physical measurements can be made. Two observers can compare their meter sticks and clocks when they are at rest to assure that they give identical results when used to measure the same quantity. When they are moving relative to one another, their measurements of time and distance will differ in accordance with the Lorentz transformation. In spite of these differences—or rather because of them—the two observers will arrive at the same description of motion. The equations describing how material bodies and electromagnetic waves move will take the same mathematical form for both observers when expressed in terms of each one's measurement of distance and time. The two observers will agree about how forces affect the motion of material bodies and they both will measure the same value for the speed of light. The laws of physics—by which we mean our mathematical description of motion—make no distinction between being at rest and moving at constant velocity. All reference frames in uniform motion are equivalent and none can make any special claim to being absolutely at rest. That is the principle of special relativity.

Observers in nonuniform motion can however presumably distinguish their absolute acceleration. Newton's second law by its very form appears to require absolute motion through space. The mass of a body times its acceleration represents a force, and the presence of that force distinguishes a body that is accelerating from one that is not. If the acceleration of a body were not absolute—if its acceleration were not uniquely determined by its absolute motion through space—then the force associated with its acceleration would not be unique and the second law would become meaningless. It would give different results for different observers depending on the reference frame that each one chose for measuring the acceleration of a body. There would be no universal description of motion.

Newton argued that the concave shape—which is actually a parabola—assumed by the surface of a liquid in a pail indicates that the pail must be spinning around its axis and that the height to which the liquid rises up the sides of the pail can be used to determine the absolute rate at which the pail is spinning. The faster it spins, the higher the liquid rises. An observer seated on a carousel feels a force directed away from the center when the carousel is rotating. The magnitude of the force depends on how fast the carousel is turning and can be used to determine the rate of rotation. The faster the rotation, the greater the force. Likewise, two objects joined by a spring and rotating about some point between them will cause the spring to stretch. An observer riding in an automobile but unable to see out would still be able to determine whether the automobile was increasing or decreasing its speed or going around a corner by simply observing the angle that a plumb bob (a weight suspended by a string) makes with the vertical.

In each of these examples, the presence of local forces indicates accelerated motion. And by the second law of motion, the magnitude of the local force determines the absolute value of the acceleration. Acceleration, it seems, must be absolute and not relative. This argument was what led Newton to believe in some absolute frame of reference with respect to which the acceleration of an object could be unambiguously determined. Otherwise, he reasoned, the second law of motion could not be valid. The fallacy in Newton's reasoning is that his argument holds only if one assumes that these forces attributed to acceleration really do result from motion. But what if such forces are not *necessarily* the result of motion? What if they can be caused by something else?

A familiar example of a force not produced by motion is gravity. Anyone who has ridden in an elevator realizes that the inertial forces resulting from the starting and stopping of the elevator would be indistinguishable from those produced by changes in gravity. When the elevator accelerates upward, one feels an additional force pushing downward, which could just as well have been caused by an increase in the strength of gravity. When the elevator decelerates and slows to a stop one momentarily feels lighter just as if the force of gravity had suddenly decreased. How can we distinguish between gravity and those forces presumably produced by accelerated motion?

We cannot, Einstein realized, and he formulated this idea as the *principle of equivalence,* which became the cornerstone of his theory of general relativity published in 1915. The result is a brand new theory of gravity in which the concept of absolute motion is no longer needed. All of the forces previously assumed to result from the absolute acceleration of mass are now attributed to gravity. Needless to say, gravity becomes much more complicated than simple Newtonian gravity. In addition, the laws of motion are written to have the same mathematical form for all observers, no matter how they are moving, whether uniformly or nonuniformly, that is, whether they are accelerating or not accelerating. In this new description of motion, no observer can lay any claim to being absolutely at rest or to having any kind of absolute motion through space. All that can be known is the motion of an object relative to other objects in the universe. Let us examine how this new description of motion is constructed.

The principle of equivalence expresses a unique property of the gravitational force: all bodies, regardless of their mass, accelerate at the same rate in a constant gravitational force. Objects in free fall move in unison, and a constant gravitational force does not produce any relative motion between them. As a result, an observer falling along with them will conclude that there are no forces acting, in agreement with the first law of motion. The only way we have of knowing about forces is through the changes they cause in the motion of bodies, and, because a constant gravitational force produces the same change in the motion of all falling bodies regardless of their mass, its presence cannot be felt or detected by an observer falling with them. Recall our earlier example of a satellite in free fall around the earth. Everything in the satellite and everything moving along with it, inside or outside, is in a condition of "weightlessness" in spite of the fact that it is all being acted upon by the earth's gravitational field. The same is true of all objects in free fall in a constant gravitational field, whether they are falling in a curved path around the earth or along some other path through space. Note too that the strength of the gravitational force doesn't make any difference. All that matters is that the force be constant, so that it accelerates all objects the same whatever the magnitude of the acceleration. All constant gravitational fields are equivalent in producing no changes in the relative motion of freely falling bodies.

This property is a rather remarkable feature of the gravitational force. The same is not true of the electromagnetic force. There the force acting on a body is proportional to both its net charge and its speed and produces an acceleration that depends on the ratio of charge to mass; in addition the acceleration is never constant but depends on the changing velocity of the body. Charge and mass are separate material properties. Bodies of identical mass can have very different amounts of charge, and the electromagnetic force does not impart the same acceleration to all bodies the way the gravitational force does.

It is in free fall in a constant gravitational field that Newton's laws of motion are observed to be valid. A body placed at rest will remain at rest, since everything falls at the same rate. A body set in motion in a given direction will appear to move in a straight line at constant speed, since in addition to its relative motion it is accelerating at the same rate and in the same direction as everything else. A force (separate from the constant gravitational force) applied to one of the falling bodies will change its motion relative to the other bodies in accordance with the second law of motion. Likewise, interactions between the falling bodies due to collisions will obey the third law of motion.

All observers in free fall find Newton's laws of motion to hold, just as they do when at rest or moving uniformly in a constant gravitational field. As a result we say that all observers in free fall in a constant gravitational force are equivalent no matter what the strength of the force, which is just another way of stating the *principle of equivalence*. The requirement that the gravitational force be constant simply means that the motion must be local, that is, limited to small distances over which the gravitational force does not change appreciably. Since each observer can carry meter sticks and clocks for measuring distance and time, the principle of equivalence states that all local measurements of motion in a reference frame in free fall are equivalent, meaning that the measurements are independent of the motion of the reference frame. Einstein assumed that the principle of equivalence applied also to the motion of light, which, as a result, should travel locally in straight paths in freely falling reference frames.

We can summarize all of this by saying that the description of motion is the same in all freely falling reference frames. Furthermore, the correct description of motion in free fall is that of special relativity, both for bodies possessing mass and for light. The theory of special relativity made the laws of physics the same in all uniformly moving reference frames. Now we have expanded the allowed motion to include uniformly accelerating reference frames. All that remains is to somehow include reference frames that are accelerating *nonuniformly*. Then the description of motion would be the same in all reference frames—for all observers—no matter how they happen to be moving, and the Newtonian concept of absolute space and absolute motion could be dispensed with once and for all. That is precisely what Einstein did in constructing the general theory of relativity.

He began with the *principle of general relativity*: the description of motion should be the same in all reference frames no matter how they are moving. To this he added the working principle that the correct description of motion in all freely falling (uniformly accelerating) reference frames is that of special relativity. To find the correct description in other reference frames, one must somehow transform the description given by special relativity from a uniformly accelerating reference frame to a reference frame undergoing arbitrary motion. And the transformation has to be done in a way that leaves the mathematical form of the description unchanged. The principle of general relativity represents the philosophical stance taken by Einstein in all of his work. What is important are not the local measurements of physical quantities such as distance and time or mass and velocity and acceleration. What is fundamental is the *mathematical form* taken by the laws of physics. Different observers may disagree about the measured values of physical quantities, but they should—and must—agree on the form taken by the mathematical equations that make up the description of motion.

Physics is a mathematical description, and it is the description itself, not the values of specific quantities that go into the description, that must be the same for all observers if we are to arrive at some common understanding of the universe. The mathematical equations by which we describe motion are the laws of nature, and if they do not have the same form for all observers, then they cannot be universal and in effect do not constitute laws at all. We would be left with no prospect of any unchanging reality behind the nature of things. Newton assumed that what was unchanging were the values of distance and time measured by all observers, no matter how they were moving anywhere in the universe. Einstein realized that it is the mathematical relationships between distance and time and not the measured values of these quantities that constitute the physical laws, and it is these fundamental relationships expressed in the equations of mathematical physics that must be the same for all observers.

That still leaves us with the practical question of how we can accomplish what is expressed philosophically by the principle of general relativity. How do we actually go about making the physical laws have the same form in all reference frames? The answer is surprisingly straightforward, if not always easy to implement. We have to reformulate the laws of physics in terms of mathematical quantities known as *tensors*. To appreciate how this works we have to briefly consider the properties of tensors.

A tensor is a mathematical quantity that can consist of multiple components or parts, each of which changes its value, or transforms, in a particular way in going from one reference frame to another. We can think of a tensor as an ordered set of quantities. The individual members of the set can be any mathematical quantities, even mathematical expressions involving combinations of quantities, which transform in the correct manner. For Q to be a tensor, by definition, its components in one reference frame have to be linear combinations of its components in any other reference frame. That means the components as determined by one observer have to be linear combinations of the components as determined by any other observer. A linear combination of the components consists of multiplying them by some set of coefficients and adding the results together. The particular set of coefficients by which they are to be multiplied is specified by the relationship, in terms of geometry and relative motion, between the two observers. The details are not important here. What matters is only this: *if the components of a tensor are all equal to zero in one reference frame, then they will be equal to zero in every reference frame*. Any linear combination of quantities, each of which is zero, will always be zero, no matter what set of coefficients we multiply them by before adding them together. This is the special property of tensors that we wish to exploit. Tensors are a sophisticated and difficult mathematical concept, and to adequately understand them requires an extensive mathematical foundation. In what follows, we will focus on only the most essential properties, those that have a direct bearing on the use of tensors in describing the motion of bodies in all reference frames, no matter how they are moving. It is for this specific purpose that tensors are employed to make the description of motion the same in all reference frames.

Suppose that we have a mathematical description of motion consisting of equations like $Q_1 = Q_2$, where Q_1 and Q_2 are both tensors. When two tensors are equal, it means that their corresponding components are equal. We can rewrite this last expression as $Q_1 - Q_2 = 0$ by subtracting the components of Q_2 from the corresponding components of Q_1 to obtain the zero tensor, the tensor whose components are all zero. (Note that in these equations 0 is not the same as the number 0; here 0 stands for the *zero tensor*, the tensor whose individual components all have the value zero.) The difference between two tensors is still a tensor. It is the tensor whose components are just the difference between the components of Q_1 and Q_2. If we let Q be the tensor representing the difference between Q_1 and Q_2, or $Q = Q_1 - Q_2$, then our description of motion takes the form $Q = 0$. Q then is equal to the zero tensor, which means that each of its components must be equal to zero. Since the components of Q are all equal to zero in one reference frame, they will be equal to zero in every reference frame. If we let Q' be the form Q takes when transformed to some other reference frame, then we will have $Q' = 0$, and our description of motion takes the same form for all observers and all reference frames. We have thus satisfied the principle of general relativity. This is the whole point of using tensors to formulate the description of motion.

This expression may look deceptively simple and devoid of content, but it is not. We have to keep in mind that Q is a tensor, an ordered set of quantities each of which can be a mathematical expression involving combinations of other quantities. The equation $Q = 0$ means that each of the components of Q must be set equal to zero, which in turn generates a set of mathematical equations, and it is this set of equations that contains the description of motion. The individual equations may, and in general will, take on very different forms in different reference frames, but the collection of equations making up the tensor equation will always have the form $Q = 0$ in every reference frame no matter what kind of relative motion is represented.

The components of Q will include things like position, velocity, and acceleration, all of which involve distance and time. Position by itself in our description of motion is not a tensor as we can easily show. Recall the Lorentz transformation of special relativity,

$$x' = \varepsilon (x - v_0 t),$$

in which x' and x represent the respective values of distance measured by observers O' and O in uniform relative motion. For position by itself to be a tensor, its component x' in one reference frame would have to depend linearly on only x, its component in another reference frame. Instead we see that x' is a linear combination of both x and t, where t is not a component of distance, and as a result the measurement of distance alone does not transform between the two observers like a tensor.

We can make a tensor, however, by combining distance and time. The quantity $Q = (x, t)$, having components x and t, is a tensor. The complete Lorentz transformation of both distance and time is given by the equations

$$x' = \varepsilon (x - v_0 t)$$

$$t' = \varepsilon \left(t - \frac{v_0 x}{c^2} \right),$$

which express x' and t' as linear combinations of x and t. Hence the quantity $Q = (x, t)$ measured by observer O transforms into the quantity $Q' = (x', t')$ measured by observer O'. The point is that to make the description of position into a tensor we have to somehow include time. The reason for this of course is that we allow bodies to move, so that their position depends explicitly on time.

In this example, we have described the position of an object by specifying its distance in only one direction from some chosen reference point. To locate an object in three-dimensional space, we have to specify its distance in each of three directions, which requires a set of three numbers whose values we can represent by the letters x, y, and z as (x, y, z). The quantity (x, y, z) is a vector, since it tells us both the distance and the direction of an object from the chosen reference point. Vectors, you will recall, are quantities that specify both a magnitude and a direction. We refer to (x, y, z) as a three-vector because it consists of three components. In Newtonian physics, three-vectors are used to specify position and to calculate velocity and acceleration. Position and time are treated as separate and independent quantities corresponding to the absolute nature of space and time.

But as we saw above, position by itself is not a tensor and does not retain the same form in all reference frames. Transforming a three-vector position to a moving reference frame introduces time and changes the spatial-only character of the vector. A description of motion using three-vectors cannot satisfy the principle of general relativity. To remedy that, we can convert our three-vector position into a tensor by including time to make it into the four-vector quantity (x, y, z, ct). Here we have included time as an equivalent distance, namely the distance ct traveled by light in a vacuum and in this way all of the components of our four-vector will have the dimensions of distance. You can think of the multiplier c, the speed of light in a vacuum, in front of the time as merely a coefficient that adjusts the time scale from seconds to the distance that light would travel in the same amount of time.

In relativity physics, we no longer treat distance and time as separate quantities but combine them into a tensor quantity that we refer to as space-time. Instead of describing motion as a succession of *positions* in a three-dimensional space, we describe it as a sequence of *events* in a four-dimensional space-time. An event is specified as a position in space at a particular time. This in a nutshell is the origin of the four-dimensional world of relativity. The fourth dimension is not something new and esoteric; it is merely time and it has been there all along. Only now we do not treat it as a separate parameter but include it as an integral part of our description of events in a four-dimensional space-time. Our four-vector position in space-time is a tensor with the correct transformation properties, and with it we can construct a description of motion satisfying the requirements of general relativity that the laws of physics should be the same for all observers no matter how they are moving.

The properties of space-time have to be consistent with the results of special relativity, in particular with the finding that meter sticks contract and clocks slow down in the direction of relative motion. Accelerating reference frames therefore introduce a distortion in the measurement of both space and time. Since the motion involves acceleration, which produces changes in both the speed and direction of motion, the distortion can vary in both magnitude and direction. Imagine for instance that we measure the circumference and the diameter of a rigid stationary circular platform and take the ratio of the circumference to the diameter. The result will be the familiar value π. Now suppose we perform the same measurement as the platform rotates about its center. The platform is accelerating when it rotates even if the speed of rotation is constant because every point on the platform is continually changing the direction of its motion. Each point in the space moves in a circle about the center of the platform. A meter stick oriented along the circumference of the platform is in relative motion (with respect to an observer stationary at its center) in the direction of its length, and hence is shortened. As a result the measured values of the circumference will be greater than before. A meter stick oriented along the diameter is not moving in the direction of its length, and hence its length is unchanged. The measured value of the diameter is unchanged. This time we do not obtain π for the ratio of the circumference to the diameter but a value larger than π. We describe the space represented by the rotating platform as *non-Euclidean* to distinguish it from the familiar Euclidean space of plane geometry. If we perform the same measurement on the surface of a sphere like the earth, we find that it too is non-Euclidean. A circle drawn on the curved surface of the earth is not a plane figure and its diameter measured along the curvature of the earth is greater than for the same circle drawn on a flat surface. This time the ratio of the circumference to the diameter has a value less than π. We speak of non-Euclidean surfaces as distorted or curved. Furthermore a clock stationary at the center of our rigid platform will run faster than one located at the rim and rotating with the platform, so that accelerating reference frames distort both space and time and cause space-time to be non-Euclidean, or curved.

These are all merely consequences of special relativity. General relativity however allows us to interpret these results in a new way. The principle of equivalence says that the effects of acceleration are indistinguishable from those of gravity, from which we can conclude that the distortion of space-time is actually due to gravity. And, since gravity is a consequence of the material property of mass, we are forced to conclude that mass and its equivalent property, energy, are what produce the curvature of space-time. It is gravity—meaning the presence of mass and energy—that slows clocks and curves space. A clock should run slower in a stronger gravitational field than in a weaker one, an effect that has now been carefully measured and confirmed to be in quantitative agreement with the predictions of general relativity.

We are also led to conclude that the motion of light is affected by gravity. An observer in free fall in a constant gravitational field will find that a beam of light travels in a straight line just as we observe it to do locally. A stationary observer not in free fall will observe that the straight path in the accelerating reference frame actually traces out a curved path in space. The acceleration of the freely falling reference frame is indistinguishable from the same motion caused by gravity, and the curved path of the light beam through space can be equally attributed to gravity. The first direct confirmation of the new theory of general relativity came in 1919 when a group of British astronomers measured the bending of starlight as it passed near the sun during a solar eclipse and found the measured deviation to be in essential agreement with that predicted by Einstein's description of the effect.

We can now sketch how we go about describing motion using the principle of general relativity. We start by describing the motion as it would appear locally to an observer in free fall or, in other words, as it would appear in a uniformly accelerating reference frame, in which the correct description of motion is that given by the theory of special relativity. We then convert the mathematical equations of the description to the appropriate tensor form by replacing all three-vectors by their corresponding four-vectors. This means correctly including time as a component of the four-vectors. It also means replacing all ordinary rates of change with respect to time by a new kind of derivative that takes into account the rate of change with respect to both space and time, or space-time. With the description expressed in terms of tensors, it will retain the same form in any reference frame, and, hence, it represents the description of motion correct in all reference frames. What can be so easy to say in words can pose formidable mathematical complexities because the resulting tensor expressions are much more cumbersome to work with than any of our previous descriptions of motion.

In addition, the procedure we have just outlined does not work when we try to apply it to our Newtonian description of gravity. Newton's gravitational equation represents instantaneous action at a distance and is completely independent of time. Since it does not contain time at all, we cannot directly convert it into a four-vector tensor form. What is needed is a different kind of description of gravity, one that is time-dependent and formulated in terms of tensors. The successful construction of such a description was Einstein's major accomplishment and secured for all time his place as the new Newton.

We have already hinted at how the new description of gravity works. The essential idea is that all accelerated motion of a body in free fall is gravitational since the principle of equivalence makes the effects of acceleration indistinguishable from those of gravity. Accelerated motion—and hence gravity—changes the measurement of distance and time in the direction of motion, shortening lengths and stretching the interval of time according to the results of special relativity. Since we are dealing with accelerated or constantly changing motion, the variations in length and time are not fixed in magnitude or direction and are not even required to be uniform but can vary nonuniformly in both magnitude and direction. Gravity is no longer the simple Newtonian force that varies as the inverse square of the distance between bodies. In the new description, gravity becomes the shape of space itself, not the ordinary three-dimensional Euclidean space but a curved, non-Euclidean, four-dimensional space-time. Gravity is still caused by the material property of mass, which is no longer constant but is velocity dependent, but now we have to also include energy to take into account the equivalence between mass and energy from special relativity. The shape of space-time is determined by the distribution of mass and energy.

Einstein made use of the already existing Riemann curvature tensor to describe the shape of space-time. Bernhard Riemann, who was a student of Carl Friedrich Gauss, had initiated the study of curved spaces of more than two dimensions as early as 1846 and had constructed the Riemann tensor to describe the curvature of non-Euclidean spaces. The curvature tensor employs a four-vector, called the *metric* of the space, that specifies the measure of distance in each of the four "directions" of space-time, corresponding to x, y, z and ct. The metric describes how space-time is distorted in each of its four dimensions. Einstein had to modify the Riemann curvature tensor to make its properties match those of another tensor already known to physicists—the stress-energy tensor that describes the distribution of mass and energy. The modified Riemann curvature tensor is referred to as the Einstein tensor.

Then, in a typically bold stroke, Einstein guessed that, since it was the presence of mass and energy that gave space-time its particular shape, the curvature tensor might simply be proportional to the stress-energy tensor, already well known from Newtonian mechanics and Maxwell's electromagnetic theory. He simply equated the two to obtain the new mathematical description of gravity,

$$G_{ij} = KT_{ij},$$

where G_{ij} is the Einstein curvature tensor, T_{ij} is the stress-energy tensor, and K is the constant of proportionality between them. The subscripts i and j are merely indices that designate the individual components of each tensor; i and j independently take on the successive integer values 1, 2, 3, 4 corresponding to the four dimensions of space-time. Each of these tensors can have components corresponding to all possible combinations of the integers 1 through 4 for i and j, namely, G_{11}, G_{12}, G_{13}, G_{14}; G_{21}, . . . , G_{24}; G_{31}, . . . , G_{34}; G_{41}, . . . , G_{44}. The tensors are 4×4 arrays consisting of four rows, each containing four components. The index i designates the row in which a component is located; j designates its position in the row, or the column in which it is located in the 4×4 array. Each component of the tensor can be a complex mathematical expression involving a combination of other mathematical quantities such as distance, time, velocity, acceleration, mass, and energy.

Einstein's gravitational equation says that each component of the curvature tensor is proportional to the corresponding component of the stress-energy tensor, so that the one tensor equation is actually a set of sixteen equations in all, all of which have to hold simultaneously and each of which can be, and in general is, quite complex. It would seem that in order to construct a description of motion that has the same form for all observers we have had to make a pact with the devil. We have traded the fairly straightforward description of Newtonian gravity for a mathematical morass that is all but intractable except in the very simplest of situations. Yet in those situations where it can be applied, Einstein's equation seems to work and provides a better description of gravitational motion than that of Newton.

The first check is to make sure that Einstein's description of gravity reduces to Newton's in the case of a static or time-independent gravitational field. This amounts to letting all velocities go to zero. In that case, Einstein's equation reduces to the familiar Newtonian expression and in addition tells us that the constant K has to have the value $K = \frac{8\pi G}{c^2}$, where G is the universal gravitational constant that appears in Newton's law of gravity and c is the constant value of the speed of light in vacuum. We have already mentioned the gravitational bending of light and slowing down of time, both of which have been measured and found to be in agreement with Einstein's equation.

In addition there is a small discrepancy in the motion of the planet Mercury when its orbit is computed using Newton's law of gravity. The planets do not really move in fixed ellipses. Instead the elliptical orbit can slowly precess around the sun, forming over time a roseate path through space. Most of the precession is due to the gravitational force exerted by the other planets. When the earth and Venus overtake Mercury in its journey around the sun their gravitational attraction drags the smaller planet along with them causing its elliptical orbit to precess in that direction. One of the major triumphs of Newton's equation of motion and law of gravity was their use in the eighteenth century to correctly compute the observed precession of the planetary orbits. But in the single instance of Mercury, in spite of repeated efforts at refinements, even including proposed modifications to Newton's law of gravity, there remained an anomaly of about 43 seconds of arc per century in the precession of the planet's orbit that could not be explained. Now 43 seconds (about 0.01 degree) of arc per century may seem at first glance absurdly insignificant; but so successful were Newton's laws of motion in computing planetary orbits that this small discrepancy was still worrisome. Einstein's law of gravity accounts for this additional amount of precession in the orbit of Mercury almost exactly, and in addition explains why the anomalous precession is not observed in the case of the other planets. This effect was cited by Einstein in his initial paper on the theory of general relativity in 1915 as a confirmation of the new description of motion.

Einstein's equation also predicts a universe that can be expanding, in agreement with observations first made by Edwin Hubble in the 1920s, which have been repeatedly confirmed and refined until today they constitute the universal view about the overall motion of the universe. What is expanding is space itself. It is not a case of matter moving outward in all directions from some center through a fixed and absolute space, as we might have pictured it in Newtonian terms. Instead every point in the universe is moving away from every other point at a rate proportional to their separation because of the uniform expansion of space itself. Whether the expansion will continue indefinitely or at some point reverse itself and become a contraction depends, according to Einstein's equation, on the total amount of matter and energy in the universe. The evidence currently suggests that there is too little matter and energy in the universe to stop the expansion, and the expansion is not slowing down but is actually speeding up and will presumably continue unchecked.

Einstein's gravitational equation also allows the possibility of regions where the concentration of mass and energy causes the curvature of space-time to become so great that no matter or energy, not even light, can escape. We refer to such regions as black holes, since their curvature effectively isolates them from the rest of the universe and makes them unobservable. Evidence for the existence of black holes in connection with collapsing stars and at the center of galaxies is accumulating, and their eventual confirmation is accepted by most astronomers and cosmologists. Some would argue that evidence of their existence is already convincing enough to constitute solid confirmation.

Unlike Newton's law of gravity, Einstein's gravitational equation is time-dependent and does not represent instantaneous action at a distance. Time-dependent variations in the distribution of mass and energy anywhere in space produce corresponding variations in the local curvature of space-time, and these variations propagate in the form of radiated energy that constitutes gravitational waves. For small disturbances far enough removed from the source, corresponding to gravitational waves that might reach the earth, the Einstein equation describes transverse waves that travel at the speed of light in vacuum like electromagnetic waves. In the case of electromagnetic waves, however, it is the electric and magnetic forces that oscillate at each point in space; whereas, during the passage of a gravity wave, what is oscillating is the local separation between two events in the four-dimensional space-time.

Space-time turns out to be a very rigid or stiff medium, meaning that it takes a very large fluctuation in mass and energy to appreciably disturb the local curvature. Put another way, very small amplitude gravity waves carry enormous energy. For the small amounts of gravitational energy expected to be bombarding the earth, the amplitude of the gravity waves involved represents oscillations in space-time comparable to the dimensions of one proton in a length of one meter. In other words, we would expect to see a meter stick change its length by only about 10^{-15} meter (that is 0.000000000000001 meter), and no one has yet succeeded in detecting motions of that amplitude. The evidence currently for the existence of gravity waves is indirect and consists chiefly of observing the orbital period of a certain binary star system. The rotation of the star system is slowing down and the slowing is in agreement with the energy that it should be radiating in the form of gravity waves as a result of its orbital motion.

Predictions made by the general theory of relativity that have been repeatedly confirmed make it seem highly likely that further evidence of black holes and gravity waves will eventually be detected. Even the failure to do so would by no means invalidate the theory, since its successes are singularly impressive and were unanticipated by any other treatment of motion. It is not wholly unusual for mathematical descriptions of physical phenomena to make predictions and yield solutions that do not correspond to anything actually observed in nature. Mathematical expressions often incorporate a flexibility that exceeds the true diversity of nature, in much the same way that language may be used to express things that go far beyond our actual experience of the world.

The rule of motion in general relativity is still Newton's equation of motion written in tensor form with four-vector forces and four-vector momentum and with the rate of change taken with respect to space-time. Under the influence of gravitational forces alone, the equation of motion predicts that a body in free fall will follow the path representing the shortest distance between any two points—any two events—in space-time. It is not necessarily the shortest distance in space nor the shortest distance in time, but the shortest distance in the two combined in space-time. Such paths are known as *geodesics*. A body in free fall moves along a path that represents the shape of space-time. In a plane in space-time, a geodesic would be a straight line since that is the shortest distance between two points but on the curved surface of a sphere in space-time a geodesic is a great circle formed by the intersection of the sphere with a plane passing through its center. In curved space-time, the form taken by geodesics depends on the exact distribution of mass and energy which determines the shape of the four-dimensional space., as embodied in the metric. We can include non-gravitational forces like the electromagnetic force in our description by putting them into the equation as four-vectors consistent with the way they are described in special relativity.

The theory of general relativity makes all motion relative and abolishes the Newtonian concepts of absolute space and time. What is absolute is not space and time but the *mathematical form* taken by the description of motion. The laws of physics are the same for all observers regardless of how they are moving. No observer by virtue of how the laws of motion appear in a particular reference frame can lay claim to being at rest in a universe in which everything else is moving. All we can ever know is how we are moving with respect to everything else. Nothing is ever absolutely at rest in the world depicted by general relativity. A world in which anything could be absolutely at rest would be a world without gravity and that in turn would mean a world devoid of mass and energy. We would then be back to the point from which we embarked with the Presocratic philosophers. The defining characteristic of our world is indeed motion, without which the universe would have to consist of nothing at all.

In the Newtonian scheme, motion is a local phenomenon that takes place in an absolute framework provided by space and time, according to a mathematical description whose form depends on the motion of the reference frame. In Einstein's scheme, space and time, along with motion, are all local concepts that obey a mathematical description whose form is identical in all reference frames. Newton's achievement first made convincing the idea that the universe obeys laws that can be expressed mathematically, that God in effect was the consummate mathematical physicist. Einstein merely extended this idea to its logical conclusion, making the laws themselves the absolute structure of the universe to which everything else including all physical measurements are relative and must be locally determined. The physical properties of the world have to vary from observer to observer in such a way that everyone is describing the same universe. Einstein's description of motion has the better claim to representing natural law since it is universal and absolute. In addition, the observational evidence is all in Einstein's favor.

The theory of general relativity revised the gravitational force to make it more like the electromagnetic force. Einstein's description makes the gravitational force time-dependent and also makes it depend on the relative velocity and acceleration between moving bodies, similar to the behavior of the electromagnetic force though the two do not exhibit the same mathematical form. An accelerating mass radiates gravitational energy in the form of gravity waves just as an accelerating charge radiates energy in the form of electromagnetic waves. Gravity waves travel at a maximum speed equal to the speed of electromagnetic waves in vacuum. Disturbances in the distribution of mass and energy at one point in the universe are not felt instantaneously at all other locations. The news propagates at the speed of light for gravity just as it does for electromagnetism. The speed of light in free space becomes the upper limit to how rapidly any part of the universe can be in touch with what is happening anywhere else.

In spite of such similarities, there are still fundamental differences between our descriptions of gravity and electromagnetism. No one has succeeded in reformulating electromagnetism to make it look like the theory of general relativity. In such a description, presumably, the electromagnetic force would be replaced by the shape of space-time that would depend somehow on the distribution of charge and electromagnetic energy as well as mass. The goal is to be able to combine the descriptions of gravity and electromagnetism into one in which both forces would be attributed to a common cause: the curvature of space-time produced by the presence of mass and charge and gravitational and electromagnetic energy. The *unified field theory,* as it came to be called, was, and still is, the holy grail of mathematical physics. The quest for it is rooted in the Greek passion for simplicity and the universal and fueled by the conviction that any God rational enough to have created a universe governed by the mathematical laws we have already discovered could not conceivably have stopped short of the perfection such a grand unifying theory would represent. What we are really talking about here is the search for divinity, and it is one that short of disillusionment and apathy we will likely never abandon. This search for order in the universe and, indeed, for the divine in the universe is after all what defines us through our consciousness.

The obstacles have proven to be formidable. For one thing, there is no equivalence principle for electromagnetism. All bodies are not accelerated equally by a constant electromagnetic force the way they are by a constant gravitational force. Instead the acceleration is proportional to the charge to mass ratio. This is not the same for all bodies, since charge and mass are separate material properties. In addition, there is only one kind of mass and two kinds of charge so that the curvature of space-time produced by negative charge would have to be in the opposite direction from that caused by positive charge. As a result, the curvature produced by a given mass would depend on whether the mass was charged positively or negatively or was electrically neutral. So far no one has discovered how to incorporate all of these features into a single, unified description without invoking a universe having more than three spatial dimensions. Likely it is simply not possible.

The general theory of relativity has become the model for how we go about formulating the physical laws so that they are the same for all observers no matter how they are moving relative to one another. The equations of mathematical physics must be written in terms of tensors, quantities that retain the same mathematical form, though not the same value, in all reference frames. Position, which by itself is not a tensor but a three-dimensional vector, is combined with time, a scalar, to produce a four-vector space-time, which is a tensor. All ordinary derivatives or rates of change are replaced by a new kind of derivative with respect to space-time. All other three-vectors—including the force in Newton's equation of motion, the electromagnetic force, and the electric and magnetic fields in Maxwell's equations—are replaced by their appropriate four-vector tensor forms. Einstein's tensor equation for the curvature of space-time replaces Newton's law of gravity. Newton's equation of motion, the electromagnetic force equation, Maxwell's equations, and Einstein's gravitational equation, all in tensor form, then constitute a complete description of motion for both matter and electromagnetic waves, one that encompasses Newton's laws of motion, electromagnetism, special relativity, gravity, and general relativity.

We might think that our task is now finished, but strangely enough that is not the case. The description that we have arrived at accurately depicts the motion of large bodies but fails miserably when we try to apply it to the motion of microscopic bits of matter or to the internal interactions of matter with electromagnetic energy. In arriving at a science of thermodynamics, heat was identified as a form of internal energy possessed by material bodies. Associated with that internal energy came an entirely new and unanticipated rule concerning the kind of processes that can occur in nature and restricting the nature of time. A similar thing happens when we try to use the physical laws that describe the macroscopic world to understand the internal motion and composition of matter. The old principles no longer work and have to be replaced by a new set of rules. The result is a radically different description of motion and along with it a fundamentally altered view of reality.

14

What You See Is What You Get

The resounding success of physics at the end of the nineteenth century in describing the motion of macroscopic bodies raises several puzzling questions when we try to apply the same description to the internal motions of matter. The laws of electricity and magnetism make it seem likely that the electromagnetic force has to play a key role in holding materials together. It is the only force that we have so far encountered strong enough to account for the remarkable cohesion of matter. All bodies contain electric charge, and even those that have no net charge contain equal amounts of positive and negative charge. The strong force of attraction between equal and opposite charges separated by small distances could easily explain the observed tensile strength of even the strongest materials. Recall from chapter 10 that Earnshaw's theorem says motionless charges cannot be held together in a stable configuration by electrostatic forces alone. Yet if we allow the charges to move, they would have to be constantly changing the direction of their motion, or accelerating, in order to remain within the confines of the material. But we learned from Maxwell's equations that an accelerating charge radiates energy in the form of electromagnetic waves.

So the mystery deepens how matter held together by electrical forces could possibly be stable. Why don't material bodies simply radiate away all their internal energy in the form of electromagnetic waves? It appears that Maxwell's equations, or at least Maxwell's equations by themselves, are inadequate to describe the internal motions of matter. And yet Maxwell's equations describe perfectly the exterior effects of electromagnetic phenomena, including the constant nature of the speed of light and the effects of special relativity.

There is also the question of how mass and charge are distributed in matter. Are they spread out over space in a continuous unbroken fashion? Or do they occur in discrete units or lumps of some smallest size, and if so how and of what are such lumps composed? We tend to think of space and time as being continuous and having no smallest unit. It is hard to think of them in any other way because we are too large and our senses too coarse to have experienced them any other way. The smallest object that we can see with our unaided eye or even with the most powerful microscope is still quite large and looks like it is made out of things smaller still. We perceive time as an unbroken, continuously flowing sequence of events. All the events we can perceive around us seem to flow continuously from events that preceded them, ad infinitum. As a result we also tend to think of physical processes as continuous even when they occur too quickly for us to observe them as a sequence or when they are too small to be seen directly. The methods of the calculus depend on the assumption of continuity and being able to break a quantity of interest into smaller and smaller pieces indefinitely. The success of the calculus in describing macroscopic motion reinforces our belief in the continuity of the physical world.

Yet, when we look about us, the world on a macroscopic scale does not appear continuous. Matter occurs in localized bodies and is not spread uniformly throughout space. To our eyes, material bodies have abrupt boundaries and in between are voids where there is no matter. Our whole way of thinking about motion is in terms of discrete material bodies that move through space as units. We even extrapolate to the concept of point masses and point charges, locations in space possessing mass and charge but no size. It might seem like a natural extension of our experience to think of matter as being composed of discrete particles rather than as a continuous distribution, but when we try to do the same thing for charge we encounter an immediate problem.

The gravitational force between particles is always a force of attraction, but the force between like charges is a force of repulsion pushing them apart. And, as the charges come closer and closer together, the force of repulsion becomes overwhelmingly strong. If we try to concentrate positive or negative charge into separate discrete units occupying a confined space, we face the question of what holds them together. Electrical forces cannot do it. It would seem that charge must be distributed continuously, so that in any finite volume of matter there is as much negative charge as positive and the net force is not one of repulsion. Either that or there must be some other force at work, one stronger than the electromagnetic force but too limited in range to be felt outside of matter.

The answers to these questions, when they came in the first decades of the twentieth century, consistently demonstrated (1) that nature is intrinsically discrete, not just the distribution of mass and charge but also the value of other physical quantities like energy and momentum; (2) that the discreteness is governed by previously undiscovered laws; and (3) that there are indeed additional forces at work in nature of which there was no inkling before the discoveries of the twentieth century.

The first piece of evidence came from efforts to describe the distribution of electromagnetic energy emitted from a small aperture in the wall of a heated cavity. The amount of energy emitted at each frequency had been measured; what remained was to explain the shape of the resulting frequency distribution by combining Maxwell's equations with the laws of thermodynamics. This seemed a perfect test of both descriptions.

According to Maxwell's equations, the enclosed cavity should be filled with electromagnetic waves radiated by the accelerating charges in the material making up the heated walls of the cavity. Waves emitted by the wall on one side travel across the cavity and are absorbed and reemitted on the other side. With the cavity maintained at a constant temperature, the laws of thermodynamics require that the electromagnetic energy in the cavity be in thermal equilibrium with the cavity walls. The heat energy emitted and absorbed by the walls of the cavity should be equal to the energy lost and gained by the electromagnetic waves in the cavity. This means that the energy carried by each wave should equal the average thermal energy of the material in the walls, which in turn is proportional to the absolute temperature of the walls.

Maxwell's equations allow an unlimited number of waves of shorter and shorter wavelengths, corresponding to higher and higher frequencies, to occupy the cavity in thermal equilibrium with the walls. Each of these waves would carry an amount of energy equal to the average thermal energy of the walls, so that the total energy in the cavity would have to be infinite. This simple but obviously absurd conclusion constituted a major embarrassment for physics in the latter part of the nineteenth century, especially since the best minds in physics could discover no error in the analysis provided by either Maxwell's equations or the laws of thermodynamics. This problem became known as the "ultraviolet catastrophe," since it predicted ever-increasing amounts of energy at higher frequencies in the cavity.

Finally in an act of desperation, Max Planck asked himself what assumptions he would have to make to force the predictions of electromagnetism and thermodynamics to agree with the distribution of energy actually emitted by the heated cavity. The correct result in this case was known from observation. Planck wanted to know what additional constraint he would have to impose on the mathematical description to come up with the right answer. Much to his surprise, and his later consternation, he discovered that the exchange of energy between the electromagnetic waves in the cavity and the charges in the walls could only take place in discrete amounts, or *quanta*, that were proportional to the frequency of the wave. The constant of proportionality, now known as Planck's constant, is customarily represented by the letter h so that a quantum of energy is given by hf, where f denotes the value of the frequency of the electromagnetic wave.

A quantum of energy is a very small amount. If we measure distance in meters, mass in kilograms, and time in seconds, the numerical value of h is 6.63×10^{-34} (that is, a decimal followed by thirty-three zeros and 663). Planck discovered that the thermal energy of the walls could only be transferred to the electromagnetic waves—and vice versa—as integer (that is, whole number) multiples of the quantum unit of energy. This rule effectively limits the energy contributed by waves of increasing frequency, since they would carry too much energy in even a single quantum to be able to exchange any energy with the cavity walls. Although Planck's new rule, when added to the description provided by Maxwell's equations and the laws of thermodynamics, produced the correct result, it appeared to be completely arbitrary and strange. Still stranger results were yet to come.

Around the turn of the twentieth century, J. J. Thomson discovered that the electrical current flowing between two electrodes in an evacuated glass tube consisted of negative charge carriers that behaved like individual, discrete particles possessing mass as well as charge. He named these particles *electrons*, and, by deflecting them in electric and magnetic fields, he was able to measure their ratio of charge to mass. A growing number of experiments suggested that the electron was the fundamental carrier of negative charge. When other experimenters introduced small quantities of various gases into these "electron tubes," they discovered that, in addition to electrons, the current also consisted of positive charge carriers. Deflecting these in electric and magnetic fields showed that they had a much smaller ratio of charge to mass, indicating that these positive charge carriers were much more massive than electrons.

Experiments by Robert Millikan conducted in the early part of the twentieth century, confirmed that charge always occurs in amounts that are an integer multiple of a fundamental quantity of charge. He measured it to be approximately 1.6×10^{-19} coulombs, named for Coulomb, whom we mentioned in chapter 10. This is the unit of charge carried by one electron and means that one coulomb of total charge consists of approximately 6×10^{18} electrons (6 followed by 18 zeros). An ampere of current, which is still a relatively small amount of current, is equivalent to one coulomb of charge flowing past a given point each second, or about 6×10^{18} electrons per second. In view of such enormous quantities of charge carriers involved in even modest amounts of charge or current, it is not hard to see why charge appears to be continuous rather than discrete on a macroscopic scale.

The fundamental unit of positive charge has the same magnitude but the opposite sign as the charge on the electron. The positive charge carrier is named the *proton*. From the measured ratio of charge to mass and the measured value of the fundamental unit of charge, the rest mass of the electron and the proton are determined to be 9.11×10^{-31} kilogram and 1.67×10^{-27} kilogram, respectively. Thus, the proton is about 1,833 times as massive as the electron. It is not surprising to find that charge occurs in discrete units of a smallest quantity of charge or that the fundamental carriers of charge are particles possessing both mass and charge, though it does nothing to explain why an electron or a proton doesn't simply fly apart from its own electrical repulsion.

Another phenomenon for which Maxwell's equations seemed to provide the wrong explanation entirely is the photo-electric effect, in which light shining on a material causes electrons to be emitted from the surface of the material. According to Maxwell's equations, the energy carried by an electromagnetic wave is proportional to the square of the amplitude of the wave, known as its *intensity*. For visible light, the intensity corresponds to its brightness. When the wave impinges on a material, this energy can be absorbed by an electron, causing it to be ejected. Therefore, the energy of the ejected electrons should also depend on the brightness of the incident light. Instead, the electron energy is found to be independent of the intensity and to depend only on the frequency, or color, of the light. Below a certain threshold frequency, which depends on the material, no matter how intense the incident light no electrons at all are emitted. Above the threshold frequency, even for an extremely faint beam of light, electrons are emitted, only fewer of them. And as the frequency of the incident light increases, the energy of the emitted electrons also increases. Below the threshold frequency, changing the intensity of the light has no effect. Above the threshold frequency, increasing the intensity of the incident light increases the number of emitted electrons or the amount of photoelectric current.

This anomalous behavior was finally explained in 1905 by none other than Albert Einstein in the very same issue of the journal that contained his first paper on special relativity. Einstein borrowed Planck's idea about the quantum of energy, only, in his typical fashion, he took it a step further. Planck had shown that energy could be exchanged between electromagnetic waves and the walls of a heated cavity only in discrete quanta of magnitude hf. But Planck's description said nothing about the distribution of electromagnetic energy in the cavity itself. Here he assumed that the energy was spread out continuously in waves as predicted by Maxwell's equations.

Einstein boldly asserted that the electromagnetic energy also propagated in the cavity in the same discrete quanta, in effect saying that light was not made up of waves after all but behaved like discrete particles—we now refer to them as *photons*—each carrying a quantity of energy equal to hf, where f is the frequency of the light. When one of these photons impinges on an electron, the photon is absorbed and its entire energy is transferred to the electron. In this process energy must be conserved, so the kinetic energy of the ejected electron is just equal to the incident photon energy minus the energy binding the electron to the material. This binding energy results from the electrical attraction between the escaping electron and the positive charge that it leaves behind in the material and has a characteristic value for each material.

In a simple equation that any student of elementary physics could derive given the same assumptions, Einstein expressed the kinetic energy of the ejected electron as

$$E = hf - W,$$

where W is the energy binding the electron to the material. Now we can see why there is a threshold frequency for emission of photoelectrons. When hf is greater than W, electrons will be emitted with a kinetic energy given by this expression. When hf is less than W, no electrons can be emitted, since they do not gain enough energy in absorbing a photon to escape the material. Millikan made careful measurements of how the kinetic energy of photoelectrons depends on the frequency of the incident light and showed that in every case Einstein's photoelectric equation correctly described the results. From his measurements, Millikan obtained an independent and more accurate determination of Planck's constant and showed that its value agreed with what Planck had found from his analysis of cavity radiation. It began to seem inescapable that electromagnetic energy occurs in nature in bundles of a minimum size, and that not only mass and charge come in fundamental units that behave like particles, but that light does as well.

Other evidence suggested that the internal energy of matter is also characterized by discrete values and is quantized. Electrons impinging on a material can cause the emission of electromagnetic radiation. The emitted radiation is observed to contain two distinct components. One is radiation distributed continuously in frequency, the kind that would be produced by freely accelerating charges not confined to specific energy values. This continuous spectrum is produced by the slowing down and energy loss of the incident electrons as they collide with the atoms in the material. The colliding electrons accelerate (change the speed and direction of their motion) and radiate electromagnetic energy in accordance with Maxwell's equations. The second component consists of radiation at only certain discrete frequencies, meaning discrete photon energies, whose values are characteristic of the emitting material. This discrete spectrum corresponds to the motion of electrons confined to distinct energy levels in the material. Electromagnetic radiation can be absorbed by the material at these same frequencies, as well as emitted. Whenever a continuous spectrum of radiation is passed through a transparent material, only certain discrete frequencies are absorbed. These absorbed frequencies, each one corresponding to a discrete photon energy *hf*, signify the presence of quantized energy levels in the material. In either case, the electrons in the material can go from one discrete energy level to another by emitting or absorbing a photon whose frequency corresponds to the energy difference.

Following the discovery of the electron and the proton, it was generally assumed that atoms were somehow made up out of combinations of these two particles. Since individual atoms are invisible it was clear that they had to be much smaller than the wavelength of visible light, which is about 10^{-6} meters (one millionth of a meter). Any object comparable to or larger than the wavelength of light can be observed by scattering light from it the way we see individual objects in the world around us. Objects much smaller than the wavelength of visible light become too small to individually scatter a light wave. In that case, a single wavelength covers a large number of atoms. The wave scatters from them as a group, so that what we see is the whole object and not the individual atoms that make it up. To see smaller objects, we need electromagnetic radiation of shorter wavelengths corresponding to higher frequencies, and hence higher energies, the kind that is produced when a beam of high energy electrons bombards a material. By scattering such short wavelength x-rays, as they are called, from crystal lattices composed of ordered arrangements of atoms, the position of the atoms could be seen as tiny dots on an x-ray photograph and their size inferred to be on the order of 10^{-10} meters.

At around the same time that Millikan was doing his research, Ernest Rutherford and his co-workers demonstrated that the positive charge in matter is concentrated in very small regions, on the order of 10^{-14} meters in size. Rutherford fired energetic beams of positively charged particles through thin foils of materials. Most of the incident particles passed through the foil undeflected, showing that the interior of the foil was largely empty space. The electrons in the foil material were too light to deflect the more massive energetic particles significantly, even in the case of a direct hit. If the heavier positive charge in the foil had been spread out in space, then it too would have deflected the incident beam through only very small angles. That was the result Rutherford expected to find. Instead a significant number of the incident particles were deflected in the backward direction, and Rutherford realized that those events had to correspond to an almost direct hit by the incident beam on a densely concentrated positive charge in the foil material. From the measured scattering angles he was able to infer that the size of the positive centers in the foil was about 10^{-14} meters, or only about 10^{-4} (one ten-thousandth) the size of an entire atom. It began to look as if atoms were composed of tiny nuclei of positive charge, presumably made up of protons, surrounded in some fashion by an equal number of electrons, with the entire arrangement being electrically neutral. Rutherford's discovery did nothing to answer the question of what could hold the positively charged nucleus together against the powerful Coulomb force of repulsion or how any such an arrangement of charges could be stable.

Soon afterward, Niels Bohr proposed a planetary model of the hydrogen atom in which a single electron orbited the positively charged nucleus, consisting of a single proton, in a circular path, like a miniature solar system held together by the electrical force of attraction instead of gravity. He described the orbits using Newton's equation of motion with the Coulomb force replacing gravity. Bohr postulated that the electron could occupy only certain stable orbits, in which the accelerating charge would not radiate electromagnetic energy as required by Maxwell's equations. The allowed orbits are those for which the momentum of the electron in its circular path, called its *angular momentum,* is an integer multiple of Planck's constant divided by 2π. Planck's constant kept popping up all over the place, each time associated with the quantization of some physical quantity, in this case the angular momentum of the atom. The energy of the electron in its orbit is related to its angular momentum, and Bohr's quantum condition amounts to a requirement that the internal energy of the atom can have only certain discrete or quantized values. With this requirement, Bohr discovered that he could correctly describe the observed frequencies of light emitted by hydrogen. When the atom goes from one allowed energy level to another, it does so by emitting or absorbing a photon whose energy is just equal to the difference between the two levels. The energy of the photon is related to its frequency by the Planck-Einstein relationship, $E = hf$.

The Bohr model of the atom works only for hydrogen. For atoms containing more than one electron, there are no general techniques available for calculating the allowed orbits and energy levels from the resulting equations of motion. But the real problem with the Bohr model is a more fundamental one. Bohr's description is based on the idea of well-defined electron orbits. In the correct description of atomic motion, the idea of precisely defined orbits and a precisely defined position for the electron has to be set aside. The electron may have a specific charge and mass and behave like a discrete particle, but all particles, it turns out, can also be described in terms of waves that are spread out in space and time.

Although it doesn't have any mass, a photon carries momentum. Light reflecting from the surface of a mirror imparts momentum to the mirror causing it to recoil. Normally we don't notice the recoil, because the small amount of momentum carried by ordinary visible light is readily absorbed by the relatively large mass of the mirror with only an imperceptible motion. According to Maxwell's equations, the momentum of an electromagnetic wave is equal to the energy of the wave divided by the speed of light in vacuum. There is a connection between energy and momentum just as there is between energy and mass.

We can understand this connection for a photon by making use of the equation relating mass and energy from special relativity, $E = mc^2$. The energy of the photon is the same as that possessed by an equivalent amount of mass, $m = \frac{E}{c^2}$. This amount of mass traveling at the speed of the photon would have a momentum equal to mc (recall that the momentum of a body is just its mass times its velocity), or $mc = \left(\frac{E}{c^2}\right)c = \frac{E}{c}$, the photon energy divided by c, just as required by Maxwell's equations. The energy of the photon can be expressed in terms of its frequency by the Planck-Einstein relationship, $E = hf$. Putting this value of E in the expression for momentum gives $\frac{hf}{c}$ for the momentum of the photon.

There is a fundamental relationship between the frequency, wavelength and speed of a wave. In each oscillation, the wave moves forward one wavelength. The frequency is the number of times the wave oscillates per second. Hence the speed of the wave is just the frequency times the wavelength (the number of oscillations each second times the distance traveled during each oscillation gives the total distance traveled each second). For a photon this relationship can be written as $c = f\lambda$, where λ represents the wavelength. Replacing c by the product of frequency and wavelength gives $\frac{hf}{c} = \frac{h}{\lambda}$ for the momentum of a photon.

Einstein's interpretation of the photoelectric effect said that light behaves as if it is actually a particle having energy hf and momentum $\frac{hf}{c}$. Yet we have given the energy and momentum of this particle of light in terms of its frequency, or equivalently its wavelength, which are wave properties that don't seem to be associated with particles at all. Our expressions for energy and momentum are inherently schizophrenic since they purport to represent the properties of a particle (albeit a particle of light) in terms of the properties that we associate with a wave, forcing the question: which is it actually, a particle or a wave? How could a particle possibly have a wavelength? What is there to oscillate in the case of a discrete particle? In addition we have the question of how a beam of particles could be bent or diffracted, the way a beam of light is, by passing it through a small opening. Einstein's interpretation of the photoelectric effect, though it explained the observed phenomena, was not very satisfying since the explanation assumed a dual nature for light. Photons are particles, but in order to know their energy and momentum we must first analyze them as waves.

If waves could behave like particles, thought Louis de Broglie, then perhaps particles could also behave like waves. He proposed that the relationship between the momentum and wavelength of a photon is a fundamental property of nature and holds for all particles, including those like the electron that have mass. The momentum, p, of the photon is equal to Planck's constant divided by its wavelength, $p = \frac{h}{\lambda}$. De Broglie suggested reinterpreting this expression to mean that every particle having a momentum p also has a wavelength $\lambda = \frac{h}{p}$ associated with it. If de Broglie's suggestion is correct, then we would expect to find that a beam of particles would behave like a wave and could be diffracted like a beam of light by passing it through an aperture about the size of its wavelength. For an electron traveling at 1 percent the speed of light the de Broglie wavelength is very small, approximately the size of a single atom. For particles of that wavelength, a suitable aperture might be the interstitial spaces between the individual atoms in a crystal lattice. Sure enough, when a beam of electrons traveling at that speed passes through a thin foil of crystalline material it is diffracted, and furthermore the deflection of the beam is in exact agreement with the wavelength proposed by de Broglie. So it looks as if de Broglie was right. All particles are somehow waves as well as particles, and the wave-particle duality of light applies to matter as well.

But now we are back to our old quandary about waves: what is actually oscillating? In the case of electromagnetic waves we said that the quantities oscillating, even in empty space, are the electric and magnetic fields, or equivalently the electromagnetic force that a charge would feel as the wave goes by. The electromagnetic wave does not represent the oscillation of anything material but describes the behavior in space and time of the force—even in empty space—exerted on any charges present. Not a very satisfying image but the one called for by the mathematical description and the best we can do. In the case of gravity waves we said that the quantity oscillating is space itself, or rather the curvature of space-time, which is determined by the distribution of matter and energy and which, as it oscillates, affects the motion of any matter present. In each case, the oscillations that make up the wave are oscillations of the propagating force, electromagnetic or gravitational, exerted on any charge or matter that we happen to put there. If we insist on concrete images, this is the picture suggested by our mathematical description and the one borne out by our observations in so far as we can test it.

What then is the nature of the wave associated with a material particle? We know of no additional force that space exerts on matter, so it seems unlikely that we are dealing with an oscillating force in this case. Does the particle itself execute some type of wave motion, or are there some kind of waves in space that guide the motion of the particle? These are questions that Erwin Schrodinger set out to investigate. He thought in terms of some type of matter waves that would guide the path of the particle. Above all he hoped to do away with the mysterious wave-particle duality by being able to clearly separate the wave itself from the particle. If the particle were a wave, then it would have to be spread out over space, and the idea of a well-defined particle having a well-defined trajectory—on which all descriptions of motion up to that point had been based—would have to be abandoned. Such a conclusion would mean that the ultimate reality behind motion is not the apparent one that we observe in the macroscopic world, and to Schrodinger such a possibility was absurd and anathema. The man who was about to correctly describe the new order was very much a champion of the old.

There is a universal equation that describes all waves. Schrodinger knew at the outset that the description of motion at the atomic level had to involve some form of the wave equation. He also knew that the kind of waves involved were characterized by the Planck-Einstein relationship between energy and frequency, $E = hf$, and by the de Broglie relationship linking momentum and wavelength, $p = \frac{h}{\lambda}$. Using these expressions for frequency and wavelength in the ordinary wave equation, and making use of the conservation of energy, we can arrive at the Schrodinger wave equation. Schrodinger attempted to interpret the solutions to this equation as some form of matter waves that directly represent the motion of a particle, but he discovered he could not do so. For one thing, the wave corresponding to a single particle moving uniformly at constant speed does not travel at the speed of the particle but moves at exactly half the particle speed. In addition, the wave itself is not localized but is spread out over all space and does not specify any precise location for the particle.

The correct interpretation of Schrodinger's waves was the work of several individuals, including Niels Bohr, Werner Heisenberg, and Max Born. The picture that gradually emerged was not one that Schrodinger found palatable. Rather than restoring the concept of a well-defined particle traveling a precise trajectory in space, Schrodinger's wave equation imposes fixed limits on how well one can know not only the location of a particle but also its speed, its total energy, its momentum and even the time during which all of these values pertain. A disillusioned Schrodinger remarked afterward that he was sorry he ever had anything to do with the new quantum mechanics, and he went off to think about problems in biology, which was and still is a completely deterministic science.

We refer to the waves described by Schrodinger's equation as wavefunctions. The term *wavefunction* merely denotes that the waves are functions of—that is, they depend on—the values of the quantities describing the particle—its mass and charge, its position and energy and momentum, and of course time. The wavefunctions do not represent the motion of the particle at all. What they give us is the expected outcome of a measurement made on the particle. The wavefunction does not tell us the exact location of the particle; it only tells us the probability that we will find it at any given position. The wavefunction also doesn't tell us anything about the location of the particle before the measurement is made or after the measurement or between measurements. It gives us nothing but the result of the measurement, and it does that only in terms of the probability of a given outcome.

The wavefunction tells us what to expect if we make repeated measurements on identical particles all having the same total energy. The result of any single measurement may differ from what the value of the wavefunction predicts. If the probability of finding the particle at some particular location is 0.5 according to the wavefunction, then if we make repeated measurements on identical particles, we should find the particle at that position about half the time. The other half it will be somewhere else. The greater the number of measurements we make, the closer we should come to finding the particle there exactly half the time. Obviously in this description of motion, the exact position of a particle cannot be specified and has no objective meaning, since all we can know about position is what we measure and the outcome of any particular measurement is specified by the wavefunction as a probability.

It isn't just the position of the particle that is uncertain, but also its momentum. A single wave with a specific wavelength corresponds to a particle moving uniformly at constant momentum given by the de Broglie relationship. But a single wave extends over all space and doesn't specify any location for the particle. We know its momentum exactly but only at the expense of having no knowledge of its position. If we wish to confine the particle to some limited region of space, we can add together a number of waves of different amplitudes and wavelengths such that the waves cancel out everywhere except in the region where we want the particle to be located. But different wavelengths correspond to different values of momentum by the de Broglie relationship. In order to confine the particle and restrict the uncertainty in its position, we have had to introduce a spread, or uncertainty, in its momentum or speed. The more tightly we wish to confine the location of the particle, the greater the number of waves required and the greater the resulting uncertainty in its momentum. To restrict the particle to a single location we have to add together literally an infinite number of waves corresponding to an infinite number of different values of momentum. The result is that we now know the position of the particle exactly, but in order to do so we have had to give up all knowledge of how fast it is moving.

This last result is an example of the uncertainty principle. From Schrodinger's equation we can show that the uncertainty in the measurement of certain pairs of quantities is always such that the product of the uncertainties (defined in a particular fashion) must be greater than or equal to Planck's constant. Expressed in symbols we can write this relationship as

$$\delta x \times \delta p \geq h,$$

where δx and δp represent the measured spread in the values of position and momentum, respectively. If we make the uncertainty in the position of the particle smaller, the uncertainty in its momentum must increase. If we try to measure the location of the particle more and more precisely so that there is no uncertainty in its position and δx approaches zero, then δp must approach infinity and we lose all knowledge of the particle's speed. If we wish to measure exactly how fast the particle is moving so that δp approaches zero, then δx must approach infinity and we can have no knowledge of where the particle is located. If we measure both the position and the momentum simultaneously, then the uncertainty in each will depend on how the measurements are performed, but the *product* of the uncertainties must always satisfy the uncertainty principle and be equal to or exceed Planck's constant.

Planck's constant is a very small number ($h = 6.63 \times 10^{-34}$) and in the macroscopic world we never notice the consequences of the uncertainty principle. Even if we were able to measure the position of a one kilogram body to within the size of a single atom, which is already far beyond the realm of possibility, then the uncertainty in its speed would only be 6.63×10^{-24} meters per second, a number so far below the range of observation as to be inconsequential if not irrelevant. But if we try to measure the position of an electron to within one atomic length, then the uncertainty in its speed is about 7.3×10^{6} meters per second, which is 2 percent the speed of light. One second after we measure its location the electron could have moved more than 7,300 kilometers (that's about 4,500 miles for those who are more familiar with distances in that unit) and we would not even know in what direction it had moved. So you can see that on an atomic level the uncertainty principle is a big deal.

Imagine being the size of an atom and trying to play in the outfield during a game of baseball. To catch a fly ball, you have to gauge both its position and how fast it is moving in order to get under it at the right time to catch it. But if you tried to determine its exact position, you wouldn't be able to tell how fast it is moving. If you tried to gauge its precise speed, then you wouldn't be able to tell where it is. The best you could hope for would be to use a big glove, try to determine the position of the ball well enough to have it land in the glove, and hope that the uncertainty principle would let you know its speed close enough to arrive under it at the right time to catch it. Even then you would never know exactly when or where it was going to hit the glove.

The uncertainty principle also applies to other pairs of physical quantities, notably energy and time. If we perform a simultaneous measurement of the total energy of a particle and the time during which it has that energy, both measurements will be uncertain by an amount consistent with the relationship

$$\delta E \times \delta t \geq h.$$

The more closely we measure the energy (δE becoming smaller) the greater the uncertainty in the time during which the particle can have that energy. Energy is still conserved over time, but the total energy can fluctuate over finite intervals of time. The shorter the interval during which the energy fluctuates (δt becoming smaller), the more the energy is allowed to vary.

To describe the behavior of a particle undergoing accelerated motion, we have to put the forces acting on the particle into the Schrodinger equation. We do this in the form of the potential energy or the work done by the force in moving the particle. The resulting equation determines the wavefunctions that describe the position of the particle and hence its motion. When we do this, we find that only certain wavefunctions corresponding to particular values of the total energy are allowed. In this way the quantum values of energy and momentum come naturally out of Schrodinger's equation and the physical constraints placed on the motion, rather than having to be imposed ad hoc the way Planck, Einstein, and Bohr found it necessary to do to obtain the right answer. With Schrodinger's equation, the quantum conditions are built right into the mathematical description itself.

For example, when we put into the Schrodinger equation the potential energy corresponding to the Coulomb force between an electron and a proton, along with the mass of both particles, we obtain the wavefunctions describing the hydrogen atom. We find that the energy of the electron and its momentum are allowed to have only certain discrete values, the same quantum values that Bohr had to assume to obtain the right answer in his analysis. But now we discover an additional constraint on the momentum of an electron, corresponding in Bohr's analysis to noncircular orbits. Now we are no longer dealing with precise orbits. Instead of following a well-defined path around the proton, the Schrodinger equation tells us that the electron is spread out over a fuzzy indistinct region whose exact shape depends on both the energy and momentum of the electron. The description provided by the Schrodinger equation reproduces the results arrived at by Bohr, but also changes them substantially and adds to them. In addition, the Schrodinger equation can be applied to atoms containing more than one electron and has been used to describe the entire periodic table of the elements, accurately accounting for the electron configurations and energy levels of all the known elements.

The allowed wavefunctions tell us all we can ever know about the motion of a physical system. Taken all together, they constitute a complete description of the system, whether it be a single particle or group of particles or an extended material body, moving under the influence of applied forces. Each individual wavefunction represents a particular configuration of the system, a particular combination of quantum values corresponding to one kind of motion the system can undergo. The total wavefunction is always a linear combination—in this case the sum—of all the individual wavefunctions and contains all of the information about the allowed motion of the system.

But the wavefunctions describe motion only by telling us the result that we should expect from making a measurement, that is, performing an observation. When we make a measurement, we find the system in some specific configuration, corresponding to the particular set of quantum values that we obtain in our measurement. Immediately following the measurement the system is no longer described by the total wavefunction but by a particular one of the individual wavefunctions. The act of measurement has selected out of all the possible configurations of the system the one that we actually find it to be in as the result of our measurement.

When we repeat the measurement on identical systems—systems consisting of identical components all arranged and prepared in the same manner—we may find the system in any one of the allowed configurations on any given measurement. The total wavefunction tells us the relative probabilities of the system being in the various configurations on any one measurement.

But what about the state of the system before we perform the measurement? When we find the system in a particular configuration as a result of a measurement, does it mean that the system was really in that configuration all along and we just didn't know it until we had performed the measurement? Our experience of the macroscopic world would lead us to believe that surely must be the case. The system doesn't know when we are going to observe it, or even if we are going to make a measurement. There is nothing obvious about the process of observation that would cause the system to change from the state it has been in all along to the one we find it in when we observe it. The system has to be in some state, and the most obvious choice would be the one we find it in as a result of observing it. But if so, what about those instances when the measurement shows it to be in some other configuration? By the same reasoning it must have been in that different state before the measurement. Out of all the possible states the system could be in, what determines its actual configuration at any given moment?

In spite of our common sense derived from experience, the best answer we can give to this question is no: before we make a measurement the system is not in a single configuration at all. It is in a state that can only be described as a linear combination of all the individual configurations in which the system can physically exist. The system somehow exists in all of its possible configurations simultaneously, and only chooses a specific state as a result of the measurement we make on it. It is this aspect of Schrodinger's equation that gives rise to the uncertainty principle. This point of view is part of the official "Copenhagen" interpretation of quantum mechanics espoused by Bohr, Heisenberg, Born, and others. It became known as that because of its close association with Niels Bohr and his theoretical physics institute in Copenhagen. It raises many questions, a number of them still not answered satisfactorily. All of these questions revolve around the issue of how the total wavefunction describing the system before a measurement "collapses" to the particular individual wavefunction that we find as a result of the measurement. The collapse of the wavefunction to give the quantum values that characterize a specific measurement has always seemed mysterious, mostly because we tend to think in terms of physical mechanisms and because no one has been able to say unequivocally what exactly constitutes a measurement. Is consciousness an essential part of measurement; and if so is a human observer necessary, or will a non-human one do just as well? Can my dog make an observation? What about an amoeba? Is measurement broad enough to include any interaction of the system with its surroundings? If so, why isn't the system always in a specific configuration? No physical system is ever completely isolated from its surroundings. These and other questions like them have been discussed and argued about since the inception of quantum mechanics in the first part of the twentieth century. They have never been completely resolved.

We have already mentioned Schrodinger's reaction to the Copenhagen interpretation. Einstein found the indeterminism of the Copenhagen view so disturbing that he abandoned the quantum theory altogether as being simply wrongheaded. He acknowledged that quantum mechanics does indeed describe the results of physical measurements correctly, but he did not believe that the Schrodinger description represented the true nature of the reality behind matter and energy. He proposed that there must be other variables in addition to the ones we ordinarily use to describe the state of a physical system, in terms of which the system at all times is in a well-defined configuration corresponding to the one we observe whenever we actually perform a measurement. The description of motion in terms of these other "hidden variables" would be completely deterministic, without the need for probabilities or an uncertainty principle. In the mid-twentieth century, the hidden variable hypothesis was championed by David Bohm, among others. Still others prefer to argue that the wavefunction has no connection to the actual state of a physical system independent of a measurement. The Schrodinger equation and the total wavefunction merely predict the outcome of a physical measurement but tell us nothing about the system at other times.

In spite of objections of Einstein and others, the Copenhagen interpretation remains the only one that is consistent with all the observations. Regardless of how bizarre it may seem to us, the evidence forces us to conclude that a system does not normally occupy a single well-defined state but coexists simultaneously in all of its possible configurations. A measurement leads always to a single result; but for the Schrodinger description to yield the right answer, we have to assume that immediately prior to the measurement all of its possible configurations are available to the system. This literally means that in order to be seen by me a single photon has to travel simultaneously along all possible paths leading from its source to my eye. As an example, a beam of light passing through a single aperture produces a very different diffraction pattern on a screen placed behind the aperture than the same beam passing through two openings side by side. If we let the light impinge on the twin apertures one photon at a time, we might think that each photon, being a single particle, would have to go through one aperture or the other, but could not go through both, and after enough photons had passed through the aperture we should get a pattern on the screen corresponding to that produced by a single aperture. Yet what we observe is the pattern produced by two apertures. Each photon somehow goes through (interacts with) both openings, or at least that description is the only one consistent with our measurement. If we try to fool the photon by using a light detector on each opening to see which one it actually goes through, then we do indeed find that each photon goes through one opening or the other, and not both, but now the diffraction pattern produced is that due to a single aperture. The measurement of which opening the photon goes through collapses the wavefunction, from that describing a photon and two apertures to that describing a photon and a single aperture, altering what we observe. The same experiment can be performed using electrons instead of photons, with identical results.

In an analysis generally referred to as Bell's theorem, John Bell suggested a way of making measurements on a quantum mechanical system that could distinguish between the Copenhagen interpretation and an alternative description based on the existence of hidden variables. The results of such sophisticated (and technically challenging) measurements performed during the past several decades have repeatedly verified that there are no hidden variables. In quantum mechanics, what you see is literally what you get and all that one can ever know with any certainty. And even that much is assured only to the extent and degree permitted by the uncertainty principle. It's what we don't, and can't, see that constitutes the ultimate reality of the microcosm. When we aren't looking, a physical system is not in any specific state but exists simultaneously in all of its possible configurations. The world of the microcosm is apparently so much more than we can ever hope to see or know.

Following the discovery of the Schrodinger equation, Paul Dirac derived a wave equation that combined the methods of quantum mechanics with the results of special relativity. The Dirac equation, as it came to be called, is the relativistic analog of the Schrodinger equation and can be used to describe particles, such as photons and high-energy electrons, moving at or near the speed of light. Dirac's point of departure was to start with the relativistic energy equation in which the total energy of a particle includes its rest mass energy (m_0c^2) so that the resulting description has built into it the equivalence of mass and energy ($E = mc^2$). By a very ingenious mathematical manipulation, Dirac was able to derive a quantum mechanical wave equation that has the same form for all uniformly moving observers who use the Lorentz transformation to compare their respective measurements of space and time. We say then that the Dirac equation is Lorentz invariant and hence satisfies all the requirements of special relativity.

The Dirac equation contained two surprising results. The first was that the particles it described had an intrinsic angular momentum of $\frac{h}{2}$, where h is Planck's constant. We call the intrinsic angular momentum of a particle its *spin*, and it had been known experimentally for some time that the electron possessed spin of $\frac{h}{2}$ but for reasons that were never clear. Now the value of the spin came directly out of the Dirac equation as a consequence of its mathematical form, which is to say its Lorentz invariance. The spin of the electron became another consequence and further evidence for the validity of the special theory of relativity.

The Dirac equation also allowed for states of negative energy in addition to the usual states of positive energy. Initially these negative energy states were not thought to correspond to anything real in nature. But in a bold hypothesis Dirac proposed that the negative energy states correspond to actual, but previously unobserved, particles. He speculated that a photon having an energy equal to at least twice the rest mass energy of an electron could disappear and in its place create two new particles, one an electron and the other a particle having the same mass as an electron but the opposite charge, called an *anti-electron*, or a *positron*. The opposite charge for the positron is necessary in order to conserve charge. Otherwise the proposed process would create net charge out of nothing. Whenever an electron subsequently combines with a positron the two annihilate and create in their place a photon having an energy at least twice the rest mass energy of an electron. This is a direct example of Einstein's relativistic energy equation relating the equivalence of mass and energy. Soon after Dirac's hypothesis, electron-positron pairs created by high energy photons were observed in cosmic ray experiments. Dirac's positron was only the first example of a new type of matter that we refer to as anti-matter. We now believe that every particle occurring in nature has its own anti-particle and many of these anti-matter particles have been created and observed in experiments performed with high energy particle accelerators. The prediction and subsequent discovery of the positron is one of the great triumphs of mathematical physics and illustrates again the utility and power of a mathematical description of motion. The form and content of the description, devised for one intended purpose, led to the discovery of totally unsuspected properties of matter and energy. The mathematics itself became the chief instrument of the inquiry. It is as if a remarkable sentence from a beautiful piece of literature written to convey one meaning was interpreted to have other possible meanings as well, and the various meanings are seen to be connected in previously unsuspected ways, giving new insights into each.

With the Dirac equation as a guide, physicists have forged a complete description of the interactions between photons and electrons in a theory known as quantum electrodynamics, or QED for short, taken from the signature conclusion of theorems in Euclid's *Geometry* (*quod erat demonstrandum*, that which was to be demonstrated). The acronym was purposely chosen to signify the completeness and precision of the theory in accounting for all known phenomena involving the electromagnetic interaction between matter and energy. QED is generally considered to be the most successful, in the sense of the most exact and comprehensive, of all physical theories to date. Its success makes the electromagnetic interaction the best understood of the four fundamental forces found in nature. The structure of QED has become the ultimate model for constructing all quantum mechanical descriptions of matter and energy.

There is much more to this part of our story than that found in this brief account. Before leaving it, we should at least briefly look at the other two forces that occur in nature, both of them associated with the nucleus of the atom and its constituents. Rutherford's discovery of the atomic nucleus left unanswered the question of what holds the nucleus together under the enormous Coulomb force of repulsion—described by Coulomb's law, which we encountered in chapter 10—between positively charged protons. The answer is that there is another force between particles in the nucleus termed the *nuclear strong force*, so named because it is the strongest of the four fundamental forces yet discovered in nature. Unlike gravity and electromagnetism, the nuclear strong force has a very short range, extending only over distances on the order of 10^{-15} meters, which is about the size of the nucleus. It is a force of attraction holding the protons together when they are confined to the nucleus but quickly becoming insignificant when they move farther apart. In addition to protons, the nucleus contains another kind of particle, the *neutron*, having approximately the same mass as the proton but no electric charge (hence its name). The nuclear strong force also exists between neutrons, and between neutrons and protons, and this additional force supplied by the neutrons helps hold the nucleus together without contributing to its Coulomb repulsion. For the lighter elements, stable nuclei contain equal numbers of protons and neutrons. For the heavier elements, more neutrons than protons are required to stabilize the nucleus against the increased Coulomb repulsion of the greater number of protons.

Finally, some unstable nuclei are observed to spontaneously emit an electron in a process called *beta decay* (before the identity of the emitted particle was known it was referred to simply as a beta ray). In beta decay, a neutron is transformed into a proton plus an electron and an elusive particle named the *neutrino* ("little neutral one") possessing no charge and a very small mass. There is an additional force termed the *nuclear weak force* acting in this process. In magnitude, this fourth fundamental force found in nature is weaker than the electromagnetic force but stronger than gravity. At this point our story branches off into the field known as particle physics, which is briefly described below. That part of the story is still evolving with the hope that new discoveries will eventually fill in the blanks, but the results so far are in good agreement with the quantum theory. The eventual outcome seems unlikely to completely overturn our conclusions thus far about the description of motion and the nature of matter and energy, though it may eventually add substantially to our understanding of them.

In fact, the success of particle physics in understanding the strong and weak nuclear forces stems from mimicking the structure of quantum electrodynamics in describing the electromagnetic force. In this description, forces between particles result from the interchange of other particles which serve as carriers of the force. The electromagnetic force between electrons results from the interchange of photons between charged particles. The photon serves as the carrier of the force. In this scheme, every force has its own carrier(s). In the so-called *standard model* of particle physics all known particles are made up of two families of more fundamental particles known as *leptons* and *quarks*. Each of these families consists of three separate "generations" of increasing mass. The electron and neutrino are the first generation of leptons, and there are two additional generations each having its own separate neutrino and a charged particle (called *mu* and *tau*) similar to the electron but of much greater mass. Thus there are six leptons in all, and all of these particles have been observed in high energy particle collisions. Protons and neutrons are composed of quarks, and the nuclear strong force is transmitted between quarks by exchange quanta called *gluons*. Quarks also come in six kinds, or "flavors," (named "up," "down," "charmed," "strange," "top," and "bottom") divided into three generations of increasing mass, and possess electric charge in units of $+\frac{2}{3}$ and $-\frac{1}{3}$ (!) the charge on the electron. The charge on quarks is the first instance of fractional units of electric charge and goes against the belief, until then, that the charge on the electron was the smallest unit of charge found in nature. A proton consists of two "up" quarks and a "down" quark (uud) to give it a total electric charge of $+\frac{2}{3}+\frac{2}{3}-\frac{1}{3}=+1$ electronic charge. A neutron consists of two "down" quarks and an "up" quark (ddu) to make it electrically neutral. Quarks also have an additional kind of charge property called "color," which comes in three types termed "red," "blue," and "green." The color property determines the strength of the interaction between quarks for the nuclear strong force the way the electric charge does for the electromagnetic force.

Gluons come in eight colors, or types, resulting from eight different combinations of the red, blue, and green color charge. Gluons play the same role in transmitting the nuclear strong force between quarks that the photon does in transmitting the electromagnetic interaction between leptons. There are important, and complicating, differences however. The photon is massless and does not carry charge. It leaves the electric charges of the two interacting particles unchanged. Since they don't carry charge, two photons cannot interact with each other. Gluons, in contrast, carry color charge and can change the color of the interacting quarks, turning one kind of quark into another. Gluons can also interact directly with each other and the theory predicts they can form bound states called *glueballs*. Since quarks carry both electric charge and color charge, they interact by means of the electromagnetic force as well as the nuclear strong force. All of this means that describing the nuclear strong interaction between quarks and between protons and neutrons is far more complicated than describing the electromagnetic force between leptons. Quarks and gluons have not been observed directly but have to be inferred from other observations.

The weak nuclear force between quarks is transmitted by three additional force carriers called the W^+, W^- and Z^0 particles, all of which have now been observed experimentally in high energy collisions produced by particle accelerators. In addition a synthesis of the electromagnetic force and the nuclear weak force into a single unified description called the *electroweak theory* has been accomplished, as a possible first step in uniting all three forces into a single theory.

In spite of its successes, the standard model of particle physics is still very much a work in progress. Many questions remain unanswered. Its successes to date are limited in the precision with which it can make predictions about such things as the strengths of interactions or the masses of particles and the quanta that serve as force carriers. It does not approach either the completeness or the precision of quantum electrodynamics, in spite of being modeled on an extension of the ideas and techniques that have worked so well in that theory. Entire phenomena, such as mass and charge, have not been completely accounted for. The theory is also unable to account for the values of the fundamental constants of nature, to specify why they have the particular values they do rather than some other values. Whether any theory should be capable of doing this or whether it is an unreasonable expectation is an unresolved question. Particle physics remains at the forefront of research in physics. Each new advance raises additional questions, with the very real possibility that none of them can be answered because of the ever-increasing requirement for higher and higher particle energies needed to pursue the answers. The ultimate answers may require energies forever beyond the reach of any man-made particle accelerator. And there is where we must leave it for now.

In the quantum description, both matter and electromagnetic energy consist of particles that propagate and interact as discrete units of mass, charge, and energy. This result is not entirely surprising given the fundamental equivalence between mass and energy uncovered by the theory of relativity. Motion is described by a wave equation, only this time the waves represent the probability of finding a particle at a given point in space and time. Position and velocity no longer possess well-defined values. Instead the location and speed of a particle obey the uncertainty principle, and what we can know about the value of either is limited by Planck's constant and what we know about the value of the other. This limitation has nothing at all to do with measurement inaccuracies but is a fundamental property of nature built right into the description of motion. The outcome of a measurement can no longer be predicted with complete certainty. Not even the state of a physical system is completely determined. The system simultaneously exists in all of its allowed configurations until a measurement collapses the total wavefunction to the particular configuration representing the outcome of the measurement.

The quantum description applies equally well to the macroscopic world, only there the extremely small size of Planck's constant makes quantum effects unobservable. For large quantum values, corresponding to the mass and energy of macroscopic objects, the description provided by Schrodinger's equation reduces to that given by Newton's laws of motion—something known as the *correspondence principle*. Quantum mechanics does not contradict our previous descriptions of motion. It merely represents the proper extension of them into the microscopic realm. The mass of a moving automobile is not continuous but is broken down into discrete particles of matter. Only the fact that each of them is so small and that there are so many of them makes it appear continuous to us. The electromagnetic signal arriving at our antenna is also not continuous but is made up of discrete bundles of energy of magnitude hf. The de Broglie wavelength of a moving automobile is a value so minuscule as to be hopelessly unobservable and totally irrelevant to our macroscopic experience of motion. The same is true concerning the uncertainty in the auto's position and speed. We simply cannot observe such small quantities. Yet according to quantum mechanics, its position and speed are still uncertain even if only by amounts that are unobservable. An automobile appears to increase its speed and hence its energy smoothly in a continuous manner as it accelerates. According to quantum mechanics, it does not do so but jumps discontinuously from one allowed quantum value of energy to the next. Only the minuscule separation between allowed energy levels—fixed by the small size of Planck's constant—prevents us from noticing it.

We find the quantum description of motion strange, even bizarre, only because we are unable to experience it directly. We cannot see what happens at or below the atomic level. Fuzzy indistinct images of the electron cloud surrounding metal atoms in a crystal lattice have been made using an electron microscope, and similar images of large molecules can be obtained, both consistent with the quantum mechanical description of matter and the restrictions imposed by the uncertainty principle. But we cannot see an electron, and when we try to describe the motion of what cannot be seen it is not surprising that we are forced to live with a description involving only what we can directly measure. Why should we be surprised to discover that the position of an electron is uncertain and cannot be completely determined? Nothing that we know about in our macroscopic world requires that it should be. We would be perhaps far more surprised to discover that motion we can see and directly observe could be equally uncertain. Yet for some kinds of motion, those described by the mathematics of deterministic chaos, that is exactly the case.

15

A Footnote on Quantum Gravity

The two most successful physical theories of the twentieth century are general relativity and quantum mechanics, pertaining to the two extremes of size and distance: from massive objects traveling at high speeds over vast distances in the theory of general relativity to the unseen motions of tiny bits of matter confined to atomic-sized distances in the quantum theory. In the intermediate realm between these two extremes, the Newtonian description of motion provides a suitable approximation to both general relativity and quantum mechanics and gives us a way—at present the only way—of bridging the gap between these seemingly disparate descriptions of reality. The two theories are formulated in completely different fashions and appear on the surface to have little or nothing in common, with no obvious way of reconciling the differences between them. General relativity is a theory of continuous space and time in which the nature of both is dependent on the distribution of mass and energy, which in turn determines the paths of moving bodies in space-time. Quantum mechanics dispenses with precisely determined paths and describes motion only in terms of the probability of obtaining a given outcome for any measurement of the position or momentum of a moving object.

It is perplexing that these two theories, each so successful in its own realm, each independently verified by countless experiments and observations and at odds with none of them, should at the same time be so fundamentally different in their approach, in their mathematical structure, and in what they appear to tell us about the nature of reality at the two extremes of size and distance. We seem to have moved away from the vaunted goal of achieving the kind of underlying unity in physical reality that the Presocratic philosophers and thinkers ever since have been searching for.

If on the smallest scale the world is quantized and made up of discrete units of matter and energy, why do such quanta not show up anywhere in our description of the world on a cosmic scale? And if the shape of space-time determines the motion of objects on a cosmic scale, why does it not enter into our description of motion on an atomic level? If both of these descriptions are correct, then surely both must be part of some more encompassing description that would reconcile the two and apply to the complete range of observations, from the very largest and most remote reaches of the universe to the tiniest and least accessible.

How to go about combining general relativity with the quantum theory remains one of the principle unresolved questions in physics at the beginning of the twenty-first century. One approach has been to look for ways of quantizing the gravitational field equation, in which the Einstein curvature tensor describing the shape of space-time is proportional to the stress-energy tensor describing the distribution of mass and energy. From the quantum theory, we know that both matter and energy are quantized and occur only in discrete amounts that are integer multiples of the fundamental quantum units. So the stress-energy tensor, to be consistent with the quantum theory, has to consist of quantized components, which means that the components of the curvature tensor must also be quantized. But the components of the curvature tensor describe the nature of space-time, implying that both space and time must somehow be quantized and can occur only in discrete amounts of some fundamental quantum units.

A theory of quantum gravity, as it has come to be called, would do away with the familiar notion of space and time as continuous, infinitely divisible quantities and replace them with a grainy space-time made up of fundamental units of a smallest size. In such a scheme, there would presumably be a lower limit to the smallest distance and the shortest interval of time that have any physical meaning in our description of motion. The distance between two points and the interval between two events would no longer decrease without limit. Space could not be shrunk beyond a minimum size; and at any point in space the interval between two events could never be shorter than a specified minimum amount. The points on a line (a real, physical line rather than the mathematical abstraction of a line that we use in plane geometry) would no longer be infinite. A real line would always consist of a finite (albeit very large) number of points or positions in space separated by a quantum unit of length between them. Likewise the hands on a real clock would not move smoothly but would jump by discrete amounts representing the quantum interval of time.

Though a theory of quantum gravity is at present only a speculative hope on the part of physicists, we can make a marriage of general relativity and quantum mechanics at least plausible by pointing to situations where both descriptions have to hold simultaneously. In the process, we can identify a set of fundamental units for measuring mass, distance, and time that occur naturally in our description of motion. These natural units suggest at least one possibility for the fundamental quantum values of these quantities.

According to the quantum theory, the energy of the universe is subject to short-term fluctuations for lengths of time that are consistent with the uncertainty principle. If δE represents the amount by which the energy of the universe varies and T represents the time during which this fluctuation occurs, then the product of δE and T must always be equal to or greater than the minimum uncertainty relationship

$$(\delta E)T = h, \text{ or } T = \frac{h}{\delta E},$$

as specified by the uncertainty principle, where h is Planck's constant. In actuality, the energy fluctuation would create a particle and its anti-particle so that within a time T the two could recombine and annihilate, returning the energy borrowed to create the mass of the two particles. This process of creating pairs of particles and their anti-particles is predicted by special relativity and quantum mechanics and is observed to occur in nature. But, for our purposes, we will suppose that the available energy of the universe increases by this amount for this length of time, and this extra energy is used to create a material particle having a mass M.

The excess energy shows up as the mass of the particle, or, according to the theory of relativity,

$$\delta E = Mc^2.$$

Using this expression in the uncertainty principle gives

$$T = \frac{h}{Mc^2}$$

as the length of time during which the particle can exist. The maximum distance in any direction R that the particle can travel in this amount of time is just

$$R = cT,$$

because, by the requirements of special relativity, nothing can travel faster than light. Combining this expression with the previous equation gives

$$R = \left(\frac{h}{Mc^2}\right)c, \text{ or } R = \frac{h}{Mc}$$

as the maximum range over which localized fluctuations of the total energy can have any physical impact on the surroundings.

So far we have used only the results of the quantum theory and the particular requirements of special relativity. But we also want to include general relativity. We can do this by requiring that the particle created by the energy fluctuation have enough mass that it becomes a black hole for anything within its range of influence, that is, for anything within a distance R. A black hole is a region in which the gravitational attraction is too strong for any form of matter or energy to escape. The amount of energy required to overcome the gravitational attraction at a distance R from a body of mass M is given by

$$\frac{GM^2}{R}.$$

If this amount of energy is greater than the total energy of the particle, then the particle will not have enough energy to escape from the region of size R and will represent a black hole. Equating this value to the total available energy of the particle gives

$$\frac{GM^2}{R} = Mc^2$$

as the basic condition that the particle represents a black hole. Solving this last expression for R and removing the common factor of M gives

$$R = \frac{GM}{c^2}$$

as the size of the black hole represented by the particle. Within this distance, we are dealing with a region where the curvature of space is too great to allow matter or energy to escape, and in this region the description of general relativity applies. Here we are dealing with black holes not on a cosmic scale but on the microscopic scale of particles created by the minute energy fluctuations allowed by the uncertainty principle of quantum theory.

We can now combine all of these results to arrive at quantitative expressions for the values of M, R, and T in terms of the fundamental constants of nature. For example, we have two expressions above for the value of R and by equating them we obtain

$$\frac{h}{Mc} = \frac{GM}{c^2}$$

from which we can solve for M to obtain

$$M = \sqrt{\frac{hc}{G}}.$$

The only quantities that appear on the right side of this equation are Planck's constant, the speed of light, and the gravitational constant. Each of these is a fundamental constant of nature whose value has been measured experimentally. Using the accepted values of these quantities ($G = 6.67 \times 10^{-11}$; $c = 2.998 \times 10^{8}$; $h = 6.63 \times 10^{-34}$) yields the value

$M = 5.5 \times 10^{-8}$ kilograms.

Substituting this value into the expression $R = \frac{h}{Mc}$, we find that

$R = 4.0 \times 10^{-35}$ meters.

From the expression $T = \frac{h}{Mc^2}$, we obtain the corresponding value

$T = 1.4 \times 10^{-43}$ seconds.

Planck was apparently the first to point out that these particular combinations of the fundamental constants h, G, and c constitute a natural set of units for expressing all other quantities of mass, distance, and time.

The Planck units represent a particular combination of these three properties for which the uncertainty principle of quantum theory and the gravitational effects described by general relativity must both hold. On the Planck scale, a local energy fluctuation permitted by the uncertainty principle could produce a black hole for which gravitational effects would have to be taken into account. Quantum fluctuations occurring within a Planck interval of time can give rise to black holes over the distance of a Planck length. The Planck mass represents the mass of such a black hole. It seems clear that, to make the two descriptions of such phenomena consistent, we must quantize general relativity, including the curvature tensor and the values of space-time.

Our analysis of the Planck units provides a straightforward rationalization for quantizing both space and time. The quantum theory reveals that the vacuum is not a static, featureless void. By virtue of the uncertainty principle, we can borrow energy from the vacuum to create pairs of particles and anti-particles that subsequently recombine to annihilate one another and return the borrowed energy to the vacuum. Variations in the energy of the vacuum state must take place within a time interval consistent with the uncertainty relationship $(\delta E)T = h$, where δE represents the borrowed energy and T the time interval over which the energy fluctuation occurs. Quantum fluctuations in the vacuum could form black holes over regions the size of a Planck length. We could not measure and hence could not have empirical knowledge of distances less than a Planck length, since a black hole is a region from which neither matter nor energy can escape. We may explore the boundaries of black holes but not the interior regions, and as a result the Planck length of 4.0×10^{-35} meters constitutes a natural dimension at which space becomes inherently discrete rather than continuous.

Space could very well be quantized on a scale much larger than a Planck length, of which we are as yet unaware and that only a complete theory of quantum gravity could reveal. But, if the quantum theory and general relativity are to hold simultaneously, then it would seem that distances less than a Planck length must be forever beyond our knowledge. Of course the exact dimension at which distance becomes discrete will most likely be imprecise and fuzzy in accordance with the description provided by general relativity and consistent with the indeterminism expressed by the quantum theory. Space-time in general relativity is not fixed and static but is constantly changing in response to the propagation of matter and energy. Gravity waves represent oscillations of space-time itself. Black holes produced by quantum fluctuations must undoubtedly exhibit the same indistinctness of size and duration demanded by the uncertainty principle.

If distance is quantized, time must be also. The only means we have of measuring time is through observations of motion such as the swinging of a pendulum or the oscillation of a spring. To measure an interval of time, we are reduced ultimately to measuring the successive positions of a moving object. The shortest interval of time possible is that required for an object to travel one quantum unit of distance at the greatest possible speed; in other words, the time required for light to travel one Planck length in vacuum, which is just one Planck interval, or 1.4×10^{-43} seconds.

Both of these values, 4.0×10^{-35} meters for the Planck length and 1.4×10^{-43} seconds for the Planck interval, are absurdly small quantities. They are so small in fact that it is impossible to even put them into a meaningful context. There is nothing in our experience to which we can compare them in order to get a meaningful sense of just how small they are. We might try to compare the Planck length to the size of the atomic nucleus for example. You may recall that we said the nucleus is only about 10^{-4} (one ten-thousandth) the size of the whole atom, and as such the nucleus represents a volume that is only 10^{-12} (one-trillionth) the total volume of the atom, which itself is already a thing small beyond our imagining. By comparison, the size of a typical nucleus is more than 10^{20} (the number 1 followed by twenty zeros) Planck lengths, a number so far beyond our grasp as to make it virtually irrelevant except as the subject of idle conversation. The best we can say for the Planck interval is that it is the time required for light to travel one Planck length. We could attempt other comparisons and physicists often try to be very creative in doing so. None of these comparisons however can do more than try to fool us into thinking that we can grasp the magnitude of quantities too far beyond our experience and our imagination for us to understand them as anything other than pure abstractions. The only real meaning that the Planck units can have for us is their role in defining the lower limits on what we are allowed to know about the universe.

There is no theory of quantum gravity at present, merely speculation and largely unsuccessful efforts to affect some kind of synthesis between general relativity and quantum theory. Some would argue that progress has been made, but as yet they cannot point to any empirical evidence in support of such claims. There is not even any compelling argument that the two descriptions should be reconcilable, beyond the abiding conviction that these two theories, each so successful in its own right, ought not to be mutually exclusive. Our analysis does nothing more than suggest the possibility that there might be phenomena in nature where both the quantum theory and general relativity are important. Because the quantum theory allows the energy of the vacuum to fluctuate does not mean that such fluctuations, even if they occur, would actually lead to black holes of the size and mass given by the Planck units. The quantum theory allows such an interpretation of what could happen, in terms of what is consistent with the limitations imposed by the theory, but that is not the same as saying that it does indeed happen. The quantum theory allows many things that it does not require. The actual world is still outside our window, not in our heads or revealed by the mathematical symbols on a piece of paper.

In the quest to synthesize these two theories, we are still captive to the Presocratic notion that there should be some single unifying reality behind the plurality of things that we observe in the world. In physics that unifying reality is motion. It is but a short step to the belief that there must be some single unifying way to describe all motion. We must always keep in mind however that the Presocratics' concept of a unifying reality grew out of a belief, and a fervent desire, that the universe should be understandable, both of which may be untenable in a world that places limits on what we can know. The real insight here, when it comes, may lie in why these two theories are not reconcilable, which in turn may lead to some deeper understanding of the reality behind things. The story of the physicist's world to this point makes such a conjecture plausible.

16

Equations That Go Berserk

Motion means change, and in a universe that is constantly changing, time is the universal parameter. The equations of mathematical physics describe how quantities such as position, velocity, and acceleration change with time. One form of change—described by equations that we term linear—has been thoroughly investigated, and as a result general techniques for dealing with such equations are available. Actually it is the other way around. Methods for solving linear equations are fairly obvious and simple, which is why they have been so widely studied and applied. All linear equations are solved by essentially the same method, or by straightforward variations of it, and so a complete mathematical understanding of this type of equation and the kind of change it describes has emerged.

Linear equations are those for which the change in some quantity is either constant in time or can at most vary at a rate that is proportional to the size of the quantity. Newton's equation of motion equates the force acting on a body to the rate of change in its momentum, which in turn is proportional to the rate of change in its position. For a constant force or one proportional to either velocity or position, the resulting equation of motion is linear. A body falling in a vacuum at the earth's surface is an example of a constant force, that due to gravity, and we have already seen the description of motion that results. An object moving at low speed through the air encounters a force of friction proportional to its velocity. For small elongations, the force exerted by a spring is proportional to the distance through which it is stretched. In each of these examples, the equation of motion is linear, yet the motion described in each case is not the same. A body in free fall accelerates at a constant rate. A falling body opposed by a frictional force proportional to its velocity will reach a terminal velocity beyond which it does not accelerate any farther. A mass on a stretched spring oscillates harmonically back and forth when released. The universal feature here is the linear nature of the equation and not the type of motion that it describes.

All linear equations have the same mathematical form, which is why a single technique can be used to solve all of them. By contrast, there are infinitely many ways in which an equation can be nonlinear, so there is no universal form and no universal method of solution. It is like describing an elephant. There is a rather narrow set of characteristics that an animal must possess in order to be an elephant, but an almost unlimited number of ways in which it can not be an elephant. Nonlinear equations must be handled individually on a case-by-case basis. A few of the simpler ones have been solved, but in general they are notoriously difficult. As a group they have been more avoided than studied, both by mathematicians and physicists.

Whenever possible, physicists use linear equations for the simple reason that they often work well enough, and when they do they are far easier to handle. Even processes known to be nonlinear oftentimes can be approximated using linear equations by making the changes involved small enough. Recently, however, it has been discovered that under certain conditions nonlinear equations exhibit a strange and unanticipated behavior referred to as *deterministic chaos*. The solutions to such equations are completely deterministic in the sense that they can be calculated from the equation, but they don't converge to a single, well-defined result. Instead they fluctuate chaotically—though not randomly—among an infinite set of nonrepeating values that depend very sensitively on the initial or starting conditions of the motion. The value of the solution at any point in time depends critically on its value at an earlier time, so that two values different by even an infinitesimal amount at one point in time may be radically different later on.

Though the solution to such an equation is deterministic, it is inherently unpredictable. It is impossible to know the initial conditions well enough to predict the outcome. To the extent that these equations actually describe physical processes, then the real world can also exhibit the same kind of deterministic but unpredictable behavior. The number of phenomena where this kind of nonlinear behavior has been observed continues to grow and includes turbulence in fluid flow, electrical oscillators, chemical reactions, pendulum motion, mechanical vibrations, celestial motion, and many others, with implications as diverse as determining the predictability (or rather the unpredictability) of the weather to determining the long-term stability of our solar system.

To get a better insight into what is meant by deterministic chaos, we want to compare the behavior of two particular equations. First we will examine a relatively simple but very prevalent linear equation, the one describing exponential growth and decay. Then, by way of comparison, we will examine one of the simpler, but nevertheless quite surprising, nonlinear modifications to this equation. Many of the relevant characteristics of nonlinear equations can be illustrated using this one example. We will let the amount of some quantity that is changing with time be represented in our discussion by the symbol A. This might be some quantity directly related to motion, such as the position, velocity, or acceleration of an object or it could be any other quantity that is changing with time, such as the number of bacteria in a culture or the amount of money in your savings account. Further, we will let A_i stand for the value that the quantity has at some instant of time, and A_{i+1} represent the value of the quantity one interval of time (1 second, 1 day, 1 year, etc.) later. During each interval of time the value of A changes by an amount that we will represent as δA. Expressing this relationship symbolically we have that

$$\delta A = A_{i+1} - A_i,$$

or in a slightly different form

$$A_{i+1} = A_i + \delta A.$$

Reading this expression it states that the value of A after the next interval of time is its value at the beginning of the interval plus the amount by which it changes during the interval.

Linear equations describe quantities that change in a particular way with time; namely, those for which the change during any interval is proportional to the value of the quantity at the beginning of the time interval. This means that

$$\delta A = cA_i$$

where c is some constant value that specifies the size of the change per unit time. For example, the increase in your savings account each year might be a certain fixed fraction (say 0.06 or 6 percent) of the amount that was in your account at the beginning of the year. In this case, $c = 0.06$ and

$$\delta A = 0.06\ A_i.$$

The total amount in your savings account at the end of the year is the amount that was there originally plus the amount by which it has increased, or

$$A_{i+1} = A_i + cA_i = (1 + c)\ A_i,$$

which is 1.06 A_i in the example that we just used.

An equation of this form,

$$A_{i+1} = (1 + c)\ A_i,$$

is said to be linear because A_{i+1} depends linearly on A_i. The amount of the quantity present at the end of each interval is directly proportional to the amount present at the beginning of the interval. If we plot corresponding values of A_{i+1} and A_i on a graph as shown in figure 16.1, the result is a straight line, hence the term linear. The larger the value of c, the more A_{i+1} increases in each interval of time and the steeper the line shown in the graph would be. There are three distinct possibilities. If $c = 0$, then $A_{i+1} = A_i$, and A does not change with time. If c is greater than zero, or positive, then

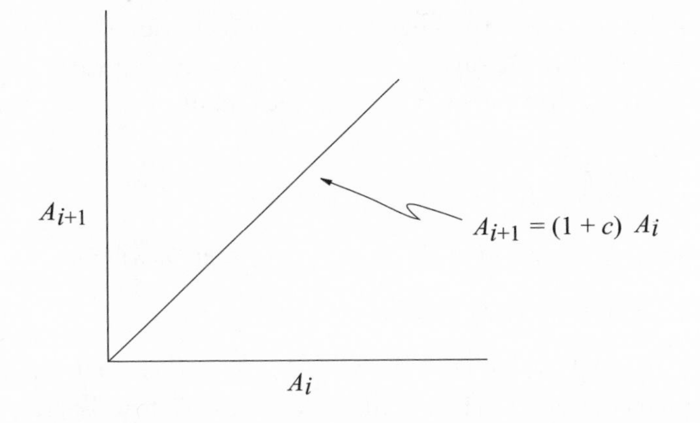

FIGURE 16.1. The relationship between A_{i+1} and A_i for a simple linear iterative equation.

A_{i+1} is greater than A_i, and A increases with time. If c is less than zero, or negative, then A_{i+1} is less than A_i, and A decreases. Keep in mind that the constant by which we multiply A_i to obtain A_{i+1} is not c, but $(1 + c)$.

The expression $A_{i+1} = (1 + c)\ A_i$ is an example of an *iterative equation.* To make use of it, an initial value is assigned to A_i. This is the value of A at the beginning of the first interval. The equation is then used to calculate A_{i+1}, the value of A at the end of the first time interval. That value then becomes the new value for A_i and is substituted back into the equation to calculate the value of A at the end of the second interval, and so on, for as many intervals as desired. Each new value of A calculated is substituted back into the right-hand side of the equation to obtain the next value of A, and the process is repeated over and over in an iterative, or repetitive, fashion, hence the term iterative equation. In this fashion, the equation traces out the behavior of A step by step in time.

Although the equation relating A_{i+1} and A_{i} is linear, the result of applying it iteratively to generate the time behavior of A is not necessarily linear. This difference between the behavior of the equation and the behavior of the solution is an important distinction. The equation that describes the time behavior of A is linear, but the variation of A itself with time may be nonlinear. One must always keep this distinction in mind in order not to misunderstand what is being said about linearity.

Let's examine how all of this works. For convenience, take the value of c to be 0.1 (a 10 percent rate of increase) so that our equation for A becomes $A_{i+1} = 1.1\ A_i$. To generate the time behavior of A, start with some initial value of A_i, say $A_0 = 1$. We call this starting value the *initial condition* on A. Then the value of A at the end of the first time interval will be

$$A_1 = 1.1\ A_0 = 1.1.$$

The value of A at the end of the second time interval will be

$$A_2 = 1.1\ A_1 = 1.1(1.1) = (1.1)^2.$$

Similarly $A_3 = 1.1\ A_2 = (1.1)^3$; $A_4 = 1.1\ A_3 = (1.1)^4$; etc. The value of A after ten intervals will be $(1.1)^{10}$; after one hundred intervals, $(1.1)^{100}$, and so on. The table below shows the successive values of A, starting with an initial value of 1 for A_0.

Successive Values of A

A_0	1	A_5	$(1.1)^5 = 1.61$	A_{10}	$(1.1)^{10} = 2.59$
A_1	1.1	A_6	$(1.1)^6 = 1.77$		. . .
A_2	$(1.1)^2 = 1.21$	A_7	$(1.1)^7 = 1.95$	A_{100}	$(1.1)^{100} = 13{,}781$
A_3	$(1.1)^3 = 1.33$	A_8	$(1.1)^8 = 2.14$	and so on.	
A_4	$(1.1)^4 = 1.46$	A_9	$(1.1)^9 = 2.35$		

If we make a graph of this table as shown in figure 16.2, we see that the result of plotting the values of A at successive time intervals is not a straight line (is not linear) but is a curve that increases slowly at first and becomes progressively steeper with each interval of time. From the table, you can see the origin of the rule that you will double your money in about seven years by investing it at a return of 10 percent interest per year. After seven time intervals, the value of A has gone from a starting value of 1 to a value of 1.95, or almost 2. It will approximately double every seven years so that in fourteen years it will have increased by almost a factor of four and in twenty-one years by a factor of seven and in twenty-eight years by a factor of fourteen, illustrating the nonlinear or accelerating pace of the increase. Note that in one hundred years every dollar invested will result in 13,781 dollars. This equation describes those quantities that

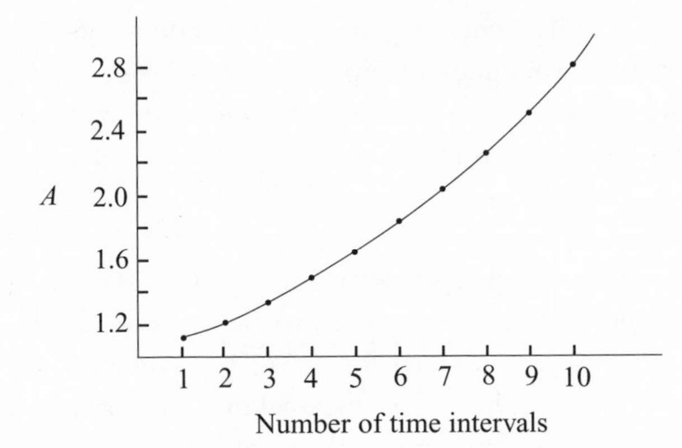

FIGURE 16.2. An example of the nonlinear output generated by the linear iterative equation shown in figure 16.1, for $c = 0.1$ and an initial value of $A_0 = 1$.

change exponentially, so designated because the number of iterations or time intervals occurs as an exponent in the result. This simple linear equation is ubiquitous in nature, describing physical phenomena as diverse as the flow of heat and the diffusion of atoms through materials, the expansion and contraction of solids, the decay of radioactive nuclei, the rate of chemical reactions, and the motion of a body acted upon by a force proportional to its velocity.

If instead of being positive c has a negative value, then A decreases with time and always approaches zero, no matter what the starting value. For example, if $c = -0.1$, then $A_{i+1} = 0.9A_i$, and starting with the value $A_0 = 1$, we obtain successive values of $A_1 = 0.9$; $A_2 = (0.9)^2 = 0.81$; $A_3 = (0.9)^3 = 0.73$; $A_4 = (0.9)^4 = 0.66$; . . . ; $A_{100} = (0.9)^{100} = 0.000027$; and so on, which approach the value zero closer and closer the more time intervals we take. Instead of exponential growth, in this case we have exponential decay. The decrease with time in the amount of a radioactive element present is an example of exponential decay; or A might represent the slowing down of a projectile acted upon by a frictional force proportional to its velocity.

It is worth noting that if the increase in A during each interval is constant rather than being proportional to A, then the equation describing A is still linear, but now A itself is also linear. If c is the constant amount by which A increases during each interval, then the value of A after n intervals is just $A_n = A_0 + nc$, where A_0 is the starting value. If c is positive, A increases linearly with time; if c is negative, A decreases linearly with time. Thus a linear equation can describe a quantity that varies either linearly or nonlinearly with time.

Next we want to examine the behavior of a nonlinear equation known as the *logistic equation*,

$$A_{i+1} = k(A_i - A_i^2),$$

where k is a constant that replaces the quantity $(1 + c)$ in the exponential growth equation. We can think of the logistic equation as a modification of the exponential growth equation in which a nonlinear term, $-A_i^2$, is included. Instead of A_{i+1} being proportional to A_i, it is proportional to the quantity $(A_i - A_i^2)$. The first term on the right side of the logistic equation is identical to the exponential growth equation and represents a change in A proportional to its value at any time. The second term represents a change in A proportional to the square of its value at any time. This latter change is always in the opposite direction to that of the first term, due to the presence of the negative sign. If k is a positive value, then the first term represents an increase in A and the second term a decrease, and vice versa if k is negative.

The logistic equation can be thought of as describing the result of competing processes, one causing a given quantity to increase the other causing it to decrease, with the rates of increase and decrease being different in value and in the way in which they change with time. This might be the exponential growth of bacteria in a culture in competition with the dying-off of the bacteria caused by limited nutrients or the buildup of toxic wastes or some other factor. It might also be the increase in a predator population, limited by the competition between predators for the available prey. Or it might be the speed of an object propelled by a force proportional to its velocity and opposed by a frictional force proportional to velocity squared. We will leave open the question of whether this equation represents anything actually taking place in the real world. For now we are only interested in its behavior.

If we make a graph of the corresponding values of A_{i+1} and A_i for the logistic equation as shown in figure 16.3, the result is not a straight line but is the curve known as a parabola, the height of which in this case depends on the value of k. In what follows we will restrict our attention to values of A_i between 0 and 1. Since we will be interested in values of k

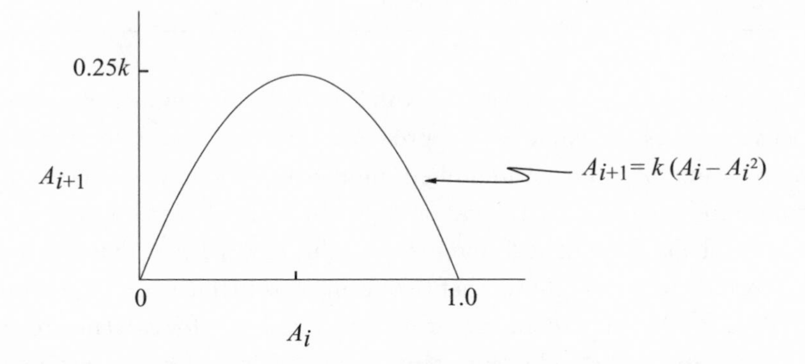

FIGURE 16.3. The relationship between A_{i+1} and A_i for the nonlinear iterative logistic equation with proportionality constant k.

less than 4, the value of A_{i+1} will also be between 0 and 1. The graph of the logistic equation shows that it is definitely nonlinear, and the feature that makes it have this nonlinear shape is the inclusion of the $-A_i^2$ term on the right-hand side of the equation.

The behavior of A described by the logistic equation depends very critically on the value of k. Recall that k takes the place of the quantity $(1 + c)$ in the exponential growth equation, where the size of c is a measure of the rate of growth or change. In the logistic equation the size of k determines both the rate of increase and decrease in the value of A, since both terms on the right side of the equation are multiplied by k. For every value of k the behavior of A is nonlinear, but as we will soon see the nonlinear behavior is far more complex than that described by the exponential growth equation. Let's examine the possible types of behavior.

If $k = 1$, the logistic equation becomes $A_{i+1} = A_i\ (1 - A_i)$, which is written in the form most convenient for computations. If we start with an initial value of $A_0 = 0.2$ for A, we find the successive values A_1, A_2, A_3, A_4, etc., given below (for this you will need a calculator):

Successive Values of *A* (*k* = 1)

A_0	0.2	A_5	0.0922	A_{10}	0.0615
A_1	0.1600	A_6	0.0837	A_{11}	0.0577
A_2	0.1344	A_7	0.0767	A_{12}	0.0544
A_3	0.1163	A_8	0.0708	and so on.	
A_4	0.1028	A_9	0.0658		

A continues to decrease toward the value zero. As the number of time intervals increases, the value of A approaches closer and closer to zero. The limiting value of A for an unlimited number of time intervals is always zero, no matter what starting value between 0 and 1 we choose for A_0. This same behavior will be found for any value of $k \leq 1$ (less than or equal to 1). A little reflection shows that this seems reasonable since $k = 1$ corresponds to $c = 0$ in the exponential growth equation; and for $c = 0$ no growth occurs and hence A should decrease with time. Values of k less than 1 correspond to negative values for c, which corresponds to exponential decay, and again A decreases with time to zero.

If k is greater than 1, say $k = 2$, then A changes with time initially but quickly reaches a limiting or equilibrium value after which it remains constant in time. For $k = 2$, the logistic equation becomes $A_{i+1} = 2A_i (1 - A_i)$, and starting with the initial value $A_0 = 0.2$ for A, we find successive values for $A_1, A_2, A_3, \ldots$, of 0.3200, 0.4352, 0.4916, 0.4999, 0.500, 0.500, 0.500, etc., after which A does not change. (You can see what happens when we use 0.500 for the value of A_i in the equation; in that case the value of A_{i+1} is also 0.500, the same as the value of A_i.) If instead we start with the initial value $A_0 = 0.7$, we obtain successive values of 0.4200, 0.4872, 0.4997, 0.500, 0.500, 0.500, etc., and the final value of A is the same as before. The final value of A is 0.500 for all initial values of A between 0 and 1 whenever $k = 2$. If we change the value of k, to say $k = 2.5$, then the final value of A also changes, in this case to $A = 0.600$, but once again A has the same final or equilibrium value for all starting values between 0 and 1.

The kind of behavior we have been describing for A up to this point—where A reaches some final equilibrium value, whether it be zero or some other number—is easy enough for us to understand by looking more closely at the logistic equation. For the value of A to be the same in two successive time intervals means that the value of A_{i+1} resulting from the equation must turn out to be the same as the value that we substituted in for A_i. For instance, if we put $A_i = 0$ in the equation, then we get $A_{i+1} = 0$, and all subsequent values of A will also be zero. Thus if A ever becomes zero it will remain zero, and as a result A cannot become negative since to do so it would first have to pass through zero. The lowest value it can assume is zero. To find the equilibrium value of A we can simply set A_{i+1} and A_i equal to one another in the logistic equation to obtain

$$A = kA(1 - A)$$

(we no longer need the subscripts since A_i and A_{i+1} are the same). Solving this equation for A gives us $A = 1 - \frac{1}{k}$. Compare this result with what we have already said about the behavior of A; notice that if $k = 1$ we obtain $A = 0$ for the final value of A; if $k = 2$, the equilibrium value is $A = 0.500$; and if $k = 2.5$ the equilibrium value is 0.600, all of which agree with what we found previously.

The logistic equation exhibits this same kind of straightforward behavior for all values of k below some critical upper limit, which is near the value $k = 3$. For k larger than this some very strange things begin to happen. For values of k immediately above $k = 3$, A does not reach a single equilibrium value but oscillates between two values. We say that the equilibrium value of A *bifurcates*. Thus if $k = 3.1$, the logistic equation becomes $A_{i+1} = 3.1\ A_i\ (1 - A_i)$; and starting with $A_0 = 0.2$, we find that A reaches the successive values 0.7646, 0.5580, 0.07646, 0.5580, 0.7646, 0.5580, etc., as you can easily verify by putting either of these two values into the equation to obtain the other. Any process actually described by this equation would not have one final equilibrium value but two possible outcomes, and in successive time intervals the quantity represented by A would oscillate back and forth between these two values. The same thing happens no matter what value of A we start with between 0 and 1, as you can demonstrate for yourself with a calculator. Similarly, if $k = 3.2$, the final values of A are found to oscillate between 0.7995 and 0.5130, as you can again demonstrate by substituting either of these two numbers into the equation for A_i. Either of these two values substituted into the equation yields the other value so that they repeat over and over. For $k = 3.3$, the final values of A are 0.8236 and 0.4794.

For slightly larger values of k an even more bizarre behavior emerges. For $k = 3.4$, A assumes not two but four final values: 0.8421, 0.4520, 0.8422, and 0.4519. You can verify this by putting each of these values in the equation $A_{i+1} = 3.4\ A_i\ (1 - A_i)$, for A_i, to obtain the next value in the sequence, which then repeats over and over. This behavior occurs no matter what initial value between 0 and 1 is used for A. If the value of k changes slightly, to $k = 3.5$ say, then we obtain a different set of four final values, in this case 0.8750, 0.3828, 0.8269, and 0.5009, which again repeat. As k continues to increase we reach a point at which there are suddenly eight final values of A, then for k even larger, sixteen values, and so on. Each time the number of values in the final sequence for A bifurcates or doubles. We call the values of k for which these successive doublings occur the bifurcation points of the nonlinear equation.

As if the behavior of this rather innocuous looking equation were not already bizarre enough, we eventually reach a critical value of k for which the equation suddenly goes completely berserk. Near $k = 3.6$ the value of A in successive intervals *never* repeats but takes on an *infinite* succession of different values between 0 and 1. We say that A has become chaotic and any process described by an equation with this kind of behavior is labeled a chaotic process. The quantity represented by A would never achieve any sort of equilibrium but would fluctuate endlessly between an infinite number of nonrepeating values.

The outcome of a chaotic process is not random, but it is not predictable either. It isn't that A can take on all possible values between 0 and 1, or even any possible set of values. It can assume only the specific values generated by the logistic equation; and, although there are an infinite number of them and no two are ever the same, they are by no means random but are completely determined by the value of k and the initial value of A used in the equation. Yet in spite of being deterministic, the chaotic values of A are inherently unpredictable. Even the slightest change in the initial value of A is enough to produce a completely different sequence of final values. No matter how close together two values of A are initially, they do not remain close together but may differ significantly in subsequent intervals; hence the expression "sensitive dependence on initial conditions" so often encountered in discussions of nonlinear phenomena. It is simply not possible to ever know the initial conditions, the starting values, well enough to predict the eventual outcome of a chaotic process. This unpredictable but nonrandom behavior is the recently discovered and unanticipated feature of nonlinear equations.

We can summarize the behavior of the logistic equation with a graph (figure 16.4) in which the final value attained by A is shown along the vertical axis and the corresponding value of k is shown along the horizontal axis. For k less than or equal to 1, the final value of A is always 0. For k greater than 1 but less than 3, A attains an equilibrium value that increases smoothly as k increases. This portion of the curve is described by the expression we arrived at previously, $A = 1 - \frac{1}{k}$. At about $k = 3$ two equilibrium values of A suddenly appear and that behavior continues until about $k = 3.4$ at which four values appear, and so on, until near $k = 3.6$ the behavior of A finally becomes chaotic and the number of values becomes infinite. In this chaotic regime, the exact sequence of values taken on by A depends sensitively on the starting value and is inherently unpredictable.

There are still further surprises in store involving the behavior of A in the chaotic regime. As we continue to increase k, we find a higher

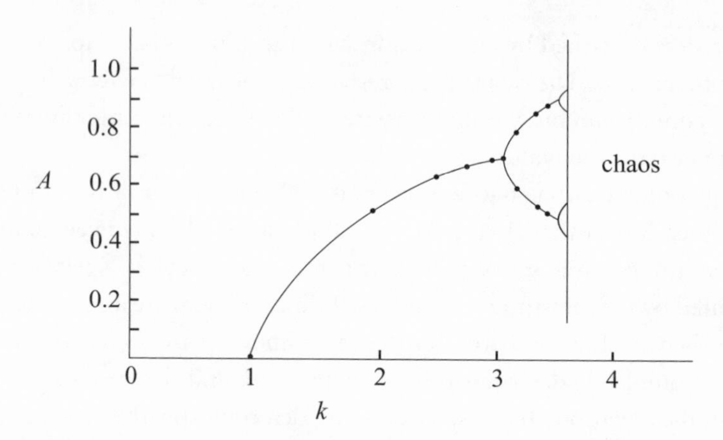

FIGURE 16.4. The equilibrium values generated by iterating the logistic equation, as a function of the proportionality constant k, showing the region of singular values, the region of bifurcation into repeating multiple values, and the onset of chaotic behavior characterized by an infinite number of nonrepeating values.

range of values for k over which the behavior of A abruptly ceases to be chaotic and once again takes on a finite number of values, only this time an odd number of final values. Each of these odd number of branches then bifurcates as before into 2, 4, 8, 16, and so on, branches to produce an even number of values of A as k continues to increase, and the behavior of A eventually becomes chaotic again for higher values of k. There are windows of order even in the chaotic regime, and these ordered regions continue to appear for higher and higher values of k. Any physical process described by an equation with this kind of behavior would be capable of making abrupt transitions from seemingly chaotic to very organized behavior.

Prior to the discovery of this strange behavior, any competent mathematician would have taken one look at the logistic equation and confidently declared that it couldn't possibly do anything very surprising. In appearance it looks simple enough, and although it is nonlinear even its nonlinear term has about the simplest and most straightforward form imaginable. More astonishing still is to discover that the behavior exhibited by the logistic equation is displayed by many other nonlinear equations, those which have a peaked or humped shape like that illustrated by the parabolic graph of the logistic equation in figure 16.3. Their behavior does not depend in any critical way on the exact relationship between A_{i+1}

and A_i and such equations exhibit common and even universal properties. Mitchell Feigenbaum was able to show that for these equations there is a universal relationship for determining the values of k at which the equilibrium value of A changes from one to two values, then four, and eight, and so on—the bifurcation points. This discovery of universality is one of the most unanticipated—and significant—properties of all. The bizarre behavior of the logistic equation is not merely a mathematical oddity or curiosity but is a general feature of certain nonlinear equations.

We can begin to understand why by depicting graphically, or geometrically, the iterative process of finding solutions to the logistic equation. The relationship between A_{i+1} and A_i expressed by the logistic equation for $k = 1$ is the parabolic curve shown in figure 16.5, where values of A_{i+1} are indicated along the vertical axis and the corresponding values of A_i are shown along the horizontal axis, as in the graph of the logistic equation that we showed earlier. Also shown is the straight line representing the linear equation $A_{i+1} = A_i$. For convenience we have restricted A_i and A_{i+1} to values between 0 and 1 as we have done all along. Notice that the parabola is everywhere below the straight line and the two intersect only at the origin, that is, at the intersection of the vertical and horizontal axes.

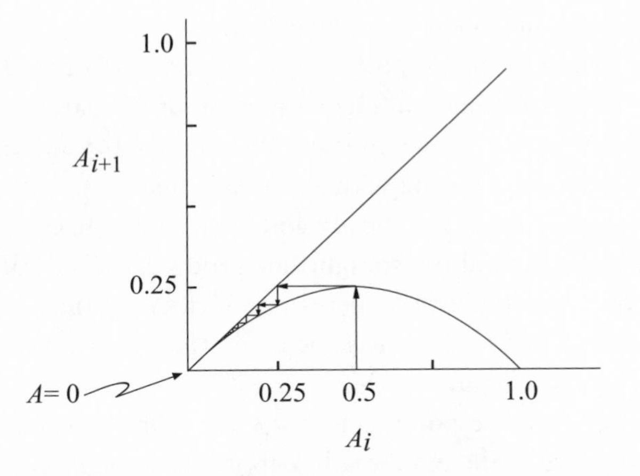

FIGURE 16.5. The process of iteration for the logistic equation with $k = 1$, illustrating how iteration leads to a final value of $A = 0$.

The iterative process of finding the final value of A can be illustrated graphically. Start with some initial value, such as $A_0 = 0.5$, along the horizontal axis as indicated in the figure. Proceed vertically as indicated by the arrows to where the arrow intersects the parabola. This point along the vertical axis represents the value of A_{i+1} corresponding to the starting value of $A_0 = 0.5$, in this case $A_1 = 0.25$. We want this to become the new value of A_i in the logistic equation to begin the next iteration. To accomplish this, we move horizontally as shown by the arrow until we come to the straight line, along which A_i and A_{i+1} are equal, which is equivalent to using the previous value of A_{i+1} as the new value of A_i. The new value of A_i is then $A_1 = 0.25$, and at this point along the horizontal axis we again proceed vertically until we intersect the parabola once more to find the next value of A_{i+1} along the vertical axis, obtaining in this case $A_2 = 0.1875$. Then we proceed horizontally to the point where we intersect the straight line (along which A_{i+1} and A_i are equal) in order to set the new value of A_i equal to the previous value of A_{i+1}. We again proceed vertically to intersect the parabola and find the next value of A_{i+1}, then horizontally to intersect the straight line to get the next value of A_i, and so on. The iterative process consists of shuttling vertically and horizontally between the parabola and the straight line as indicated by the arrows in figure 16.5, following the arrows until they lead us to the final value of A if one exists. From the figure it is clear that the result in this case will be a final value of $A = 0$, where the parabola and the straight line intersect at the origin.

The next graph (figure 16.6) shows the parabola corresponding to $k = 2$ in the logistic equation. Here a portion of the parabola lies above the straight line, and the two intersect at a value of 0.5 along the vertical and horizontal axes. Starting with an initial value of $A_0 = 0.1$, as shown, and following the arrows vertically and horizontally back and forth between the parabola and the straight line produces a final value for A of 0.5, the point where the two curves intersect. Any other starting point between 0 and 1 will lead to the same final value for A as you can readily determine from the figure.

The parabola corresponding to $k = 3.25$ (figure 16.7) is shown in the next figure. More of the parabola lies above the straight line in this case and the two intersect at a value of $\frac{2.25}{3.25}$. By following the arrows we see

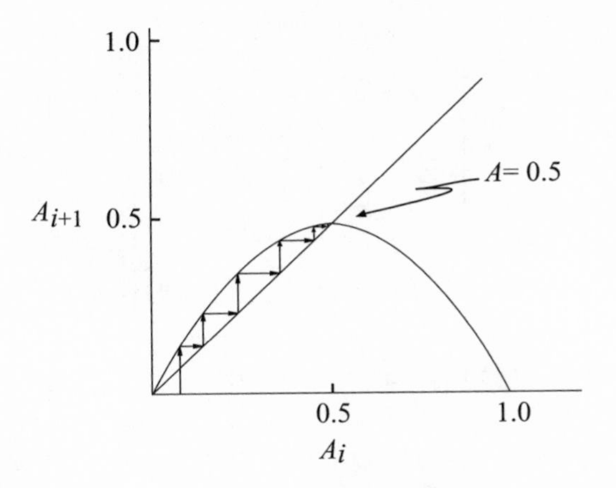

FIGURE 16.6. The process of iteration for the logistic equation with $k = 2$, illustrating how iteration leads to a final value of $A = 0.5$.

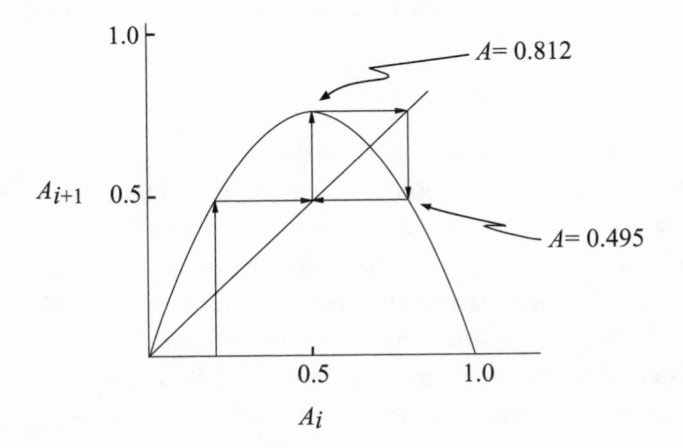

FIGURE 16.7. The process of iteration for the logistic equation with $k = 3.25$, illustrating how iteration leads to bifurcation, with repeated values of 0.812 and 0.495.

that now there are two final values of A that repeat over and over sequentially as the iterative process is continued. Finally, the parabola corresponding to $k = 4$ shown in figure 16.8 illustrates what happens in the chaotic regime. Following the arrows reveals that no value of A ever repeats, and hence A now takes on an infinite number of values.

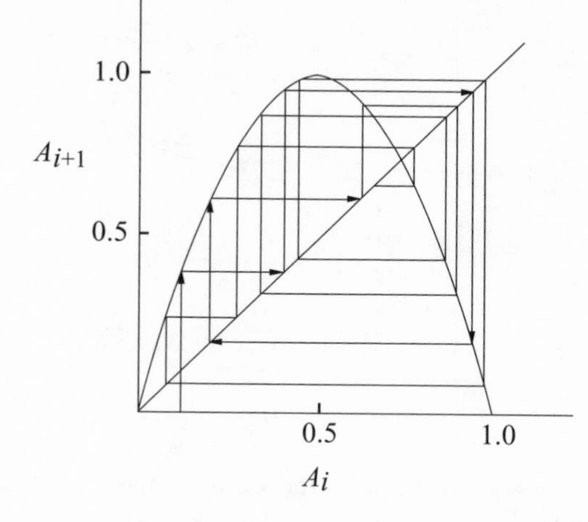

FIGURE 16.8. The process of iteration for the logistic equation with $k = 4$, illustrating how iteration leads to chaotic behavior with an infinite number of nonrepeating values.

From this graphical depiction of the behavior of the logistic equation, we can see why other nonlinear equations exhibit similar behavior. The behavior does not depend on the exact shape of the curve relating A_{i+1} and A_i. The same type of behavior can occur for other peaked or humped curves where part of the curve lies above the straight line. The numerical value of k is critical in this example because it adjusts how much of the curve representing the logistic equation falls above the straight line and where the two intersect. It was this universal feature of nonlinear equations that Feigenbaum discovered.

What implications does any of this have for our mathematical description of motion? Newton's equation of motion can be linear or nonlinear, depending on the form taken by the mathematical expression describing the force acting on a moving body. For a force that is directly proportional to the displacement of an object from some point, the equation of motion is linear and has only stable solutions. In this case the equation of motion is identical in form to the wave equation and describes the same type of motion—that of oscillations back and forth around a position of stable equilibrium. The displacement of the object as it oscillates traces out a wave in time. There is a very general mathematical theorem, known as Taylor's theorem, which says that for sufficiently small displacements away from stable equilibrium any force is approximately proportional to the displacement. The equation of motion therefore is always linear for an object or a physical system making small enough excursions around a position of stable equilibrium. As the displacement away from equilibrium increases, the force becomes increasingly nonlinear and depends on the square or the cube or higher powers of the displacement. The resulting motion can then have the kind of unstable and even chaotic behavior exhibited by the logistic equation.

The wave equation is linear but only for waves of small amplitude or for waves propagating through media having constant or linear properties. The wave equation for electromagnetic waves in vacuum for example is strictly linear. But all real materials have nonlinear properties that depend on the amplitude of the wave and for them the wave equation becomes nonlinear. We still often use the linear wave equation as an approximation that is sufficiently accurate for waves of small enough amplitude, but a variety of nonlinear effects accompany the propagation of matter waves and electromagnetic waves through real materials.

Newton's law of gravity, with its force that varies as the inverse square of the distance between two masses, makes the resulting equation of motion nonlinear. The motion of two bodies, such as one planet revolving around the sun, is stable and well-behaved. But, for any three bodies moving under the influence of gravity, the eventual motion is unpredictable due to its sensitive dependence on the initial conditions. Knowledge of this unpredictability has generated renewed interest in the long-term stability of the solar system.

The equations that describe the turbulent motion of fluids are inherently nonlinear, and for them linear approximations are of limited utility. Turbulence is intrinsically chaotic, and the motion is unpredictable due, among other things, to the impossibility of ever knowing the initial conditions accurately enough. The earth's atmosphere and its weather patterns represent an example of turbulent fluid flow. Long-range weather forecasting is not a matter of bigger and faster computers and better mathematical models. The models themselves are nonlinear, and we now know they can never be used to make exact predictions.

These are just a few of the implications resulting from the strange behavior of nonlinear equations. There may be other implications as well. The windows of order that can suddenly appear in the midst of chaos are the origin of the ordered structure that we observe in even chaotic phenomena like turbulence. Ilya Prigogine has suggested that the origin of life may turn out to have been such a window of order arising out of the atomic and molecular chaos of the physical world around us. The logistic equation becomes unstable and then chaotic as the parameter k increases. The value of k determines the rates of increase and decrease that compete with one another in the equation to determine the final value of A. As k increases, the equation describes a quantity pushed farther and farther away from a happy balance between increase and decrease. Nonlinear systems tend to become chaotic when forced too far away from a stable equilibrium, yet forced farther away they can suddenly become ordered once more. Then the order is a precarious one kept from being chaotic by only a narrow margin between all the competing influences that force the behavior of the system in its various directions. In the narrow windows of order arising out of chaos is the hint that there may be rules or principles of organization representing new physical laws of nature of which we are as yet unaware, principles that pertain not to the motion and interactions of individual atoms and molecules but to interactions between large aggregates of particles in the form of collective phenomena. The behavior of genes and other self-organizing systems comes to mind, as does the fundamental question of how such self-replicating systems ever arose in the first place. This will surely be an area of intense interest and active inquiry in the future chapters of our story, although the problem of trying to discover such principles of organization is undoubtedly going to be extremely difficult given the complex nature of a nonlinear mathematical description.

For now the chief consequence for our description of motion is the discovery that even in the macroscopic world there are phenomena that we can describe mathematically without being able to know with any certainty what the future course will be. Even in the realm of our immediate sensory experiences, in the world that we can see and touch, nature limits what we can know, not as a consequence of the imprecision of our senses or the inaccuracies of our instruments but once again because of the very nature of the world itself.

17

The Physicist's World

In trying to fathom the universe, we are prisoners of our own irony. With consciousness comes awareness, but with awareness comes also the realization that it is after all only a limited awareness, one that leaves us always with more questions. Being aware of the questions, we naturally want, and expect, answers. That we could be aware at all suggests there must be answers. Yet any child can ask questions about the nature of the world that can never be answered absolutely, only conditionally. The conditions may be grounded in belief as often as in reason and may be expressed as convincingly by mythology or religion as by science. Incapable of answering *why* things happen as they do, and thinking to escape the frustration of that limitation, the physicist turns instead to describing *how* things happen, only to discover that the description comes with its own limitations.

By describing how things happen, we hope to understand how they came to be and to that limited extent to satisfy our most basic desire to explain why the world is the way it is and not some other way. Our hopes have foundered on the discovery in the twentieth century that a complete descriptive science is no more possible than the explanation behind things that it was intended to replace. What makes anything happen at all of course is some type of motion; physics, therefore, is a description of motion. And that is all it is, though at times our ability to describe how material bodies move may seem to give us an understanding that goes beyond the simple description of motion and includes all of the other attributes of the universe. We developed descriptions of other attributes, things like mass and charge, and leptons and quarks, and esoteric properties such as spin and color charge, only because we have found them necessary to be able to describe our observations of how material bodies and energy move. Until the twentieth century, it was believed that the description could be made exact. The limits to knowledge were thought to be merely practical, having to do only with the present incompleteness and imprecision of the measurements involved in human observations. Now we know that, in addition to such practical limits, there are fundamental constraints having nothing to do with human limitations but deriving from the nature of reality itself.

We may want to believe, with Newton, in the existence of an absolute and universal space and time. In this view, space is the fixed framework through which all material bodies move, the same everywhere, and time is the parameter by which all observers will record the successive positions of a moving body as it travels through this fixed space. But such a view requires that we be able to demonstrate that any two observers, no matter what the state of their motion, will agree on their respective measurements of space and time. Otherwise such concepts can have no claim to being absolute and universal.

Here we run head on into what is the most fundamental observation of all about the universe, namely that all observers in uniform relative motion measure the same value for the speed of light in vacuum (that is, in the absence of any material bodies). This finding, in turn, requires that uniformly moving observers must differ in their measurement of both space and time in the manner described by the Lorentz transformation. Otherwise the two observers would measure different values for the speed of light. Distances contract in the direction of motion and moving clocks run slower, in precisely the manner required to make the speed of light the same for each observer. We have to set aside the idea of an absolute space and time and, along with it, any idea of absolute motion. We cannot answer the question of what is at rest and what is moving, only the question of how one body moves relative to another. The most we can say is that nothing in the universe can be known to be at rest. And since nothing can be known to be at rest, then nothing is at rest. What is universal and absolute in the description of motion is not the measurement of space and time but the mathematical form of the description itself, the form taken by the equations out of which we construct the physicist's world. The laws of physics must have the same form for all observers no matter how they are moving. That is the crux of the matter and the one principle on which all of our descriptive understanding of the universe is based.

We must always keep in mind too that the physicist's world is constructed wholly out of quantities that can be measured, and it is in that sense that it is a physical edifice as well as a mental one. All that we can know about the world is what we can measure—things like position, time, velocity, acceleration, force, mass, and charge. This is already a very limited kind of knowledge but it is the only kind that concerns us here. If we want to know a property that something has, we must be able to state how we measure it. If we want to know whether we are both talking about the same thing, we must both measure it and compare our results. It does no good to ask at the same time: Yes, I know this is how one measures it, but what is it really? It does no good to look for some other reality behind the appearances, beyond what we can actually observe. In the physicist's world, reality and appearance are not separate and distinct but are one and the same: what can be measured. Any distinction between the two belongs to the realm of philosophy, not physics. We may speculate endlessly about what cannot be directly observed, but, until we can actually measure the physical properties of electrons or protons, we can never be assured of their existence. We may choose to push our mathematical description of motion far beyond the limits of observation and measurement, as in trying to describe the interior of a black hole or the internal makeup of the elementary particles, but so long as we do, our picture of things remains speculative and certainly cannot be regarded as either convincing or unique. And in such cases we can claim no real knowledge of the world.

The only quantities in the universe that can be considered absolute are those for which all observers obtain the same result. For observers in uniform motion the speed of light in a vacuum is absolute. Distance and time are not. Neither is mass, since its value depends on the relative speed of the body with respect to the observer. Charge is absolute but charge density (the amount of charge per unit volume) is not, since distance (and hence volume) contracts in the direction of motion. All observers, no matter how they are moving, can determine the mathematical relationships that exist between the measured quantities used to describe motion, and when they do they should all obtain the same result. The laws of motion themselves are absolute.

Einstein adopted this principle as the fundamental postulate on which he based the theory of relativity and suggested it as the proper foundation for all of physics. The surprise is not that he discovered its usefulness, or that he was able to formulate a description of motion for which it is valid, but that it went so long unarticulated. It amounts to little more than saying that we are all describing the same universe, but its adoption in the first part of the twentieth century represented a distinct shift in our point of view. Before then we thought that the measurement of physical quantities like distance and time and mass must be preserved. Now we realize that it is not the measured quantities but the form taken by the mathematical relationships between these quantities that is sacrosanct. Prior to Einstein's enunciation of the principle of general relativity, we were seeking simplicity in the wrong place. Before Kepler discovered that the planets move in elliptical paths around the sun, simplicity in the motion of the heavenly bodies meant orbits that could be described using only circles. This turned out to be a false simplicity that necessitated a complex system of "epicycles" to match what was observed with the senses. Newton revealed that bodies in the gravitational field could move along any of the conic sections, depending on the total energy of the motion. The true simplification lay not in the simplifying properties of circles, as the Greeks supposed, but in the simplifying realization that all of the allowed paths of motion have a common mathematical form and origin.

Since the laws of physics are independent of the motion of the observer, we cannot regard anything in the universe as being at rest. There are no measurements by which we can determine anything other than the relative motion between two bodies. Everything moves, and nothing is ever absolutely at rest. Even two bodies that appear to be at rest with respect to one another are not really so. If we are to accept the quantum theory as a correct description of atomic motion, then at the microscopic level nothing can be motionless. We can confine a particle of matter to a single position at one instant of time only by relinquishing all knowledge of where it will be the next instant. As the uncertainty in its location decreases toward zero, the uncertainty in its momentum must increase without limit in accordance with the uncertainty principle. We are not talking about an uncertainty in its position due to some failure of human perception, but an inescapable bound on our ability to measure, and hence to know, its position. Quantum mechanics tells us that even at a temperature of absolute zero bodies still exhibit internal atomic and molecular motion. Everywhere in our description of the world, we find vindication of Heraclitus's assertion that everything changes and nothing abides, that you cannot step into the same river twice, or even once for that matter. The most fundamental feature of our world is motion, and we cannot consider the nature of matter or energy independently of how things move. Matter and motion are not separable attributes of the world; the properties of matter, and those of energy, derive entirely from motion. We can have no knowledge of one without the other.

The absolute nature of the speed of light in vacuum leads us to the Lorentz transformation as the description of how two uniformly moving observers compare their respective measurements of distance and time. The Lorentz transformation in turn fixes the speed of light in vacuum as the upper limit for any attainable physical speed. This result, though somewhat startling when first encountered, should not be surprising. If there were no upper limit to how fast mass and energy could be transmitted from one place to another, then all parts of the universe would be instantaneously in contact. Anything occurring anywhere in the universe could be known everywhere else as it happened, with no delay. Each event would define a universal and absolute Now, and Newton's concept of absolute time would be realized. The farthest reaches of the universe would be instantly accessible from any place within it, and the whole universe would in essence be shrunk to a single place or point in space and time. Distance would become irrelevant and in effect the universe would no longer be three dimensional but one dimensional.

As it is, the finite value of the speed of light has the effect of isolating us from all but a tiny portion of our surroundings. In one lifetime, we can have direct knowledge of what occurs only within a few tens of light years from us. In the time that humans have existed, news of events from outside our own galaxy—roughly one hundred thousand light years in diameter—would not yet have reached us. Everything we see happening we see in the past, not in the present. For us there literally is no present, only a past indicated by the motion of matter and energy. When we look at the night sky, we see it as it once was, when the light first began its journey to our eyes. It is questionable whether humans will survive long enough to be certain of anything outside our own immediate surroundings, or to know for sure if even our own galaxy still exists in the form that we now observe. Edwin Hubble expanded our concept of the universe from one in which the Milky Way was believed to be the only galaxy to one that included maybe one hundred billion other galaxies and distances of perhaps thirteen or fourteen billion light years. He did so merely by peering farther into the past. Recently the light from a powerful pulsar—some kind of vast cosmic explosion with an energy output exceeding that of entire galaxies like our own—was observed by astronomers who put it twelve billion light years away, meaning that it took place twelve billion years ago and it has taken that long for the light to reach us. If such an explosion took place in our own galaxy, we might not know about it for tens of thousands of years. All that we can say for certain is that such an explosion could not have occurred much before then.

Causality provides us the only link we have with the rest of the universe outside of our local surroundings. In trying to infer the nature of the universe now from observations of events that all occurred in the past, it becomes crucial whether causality is a permanent and unchanging property of the world. Do things still happen as they always have, or is the causal relationship between physical phenomena one that has changed during the evolution of the universe? Are the laws of nature themselves (by which we mean our description of the motion of material bodies and energy) really absolute, or do even they evolve along with the rest of the universe? When we look back into the past at the universe as it once existed, are we justified in assuming that subsequent events in our entire surroundings have always occurred as we observe them locally at the present time? We begin to realize just how important and how devastating to physics was David Hume's skepticism concerning the necessity of causality, and we are forced to conclude that we do not, and cannot, know the answer to such questions. Our knowledge of the world is all the more local and all the more problematic and conditional as a result.

Immanuel Kant's reply to Hume notwithstanding, no one has succeeded in answering Hume's objections because they are unanswerable and represent one of the fundamental limitations to human knowledge. Yet our entire understanding of the universe is based on the assumption of causality. When we view our surroundings, we see galaxies that exhibit common shapes and common properties though separated by millions and billions of light years. Similar galaxies exhibit similar behavior,

behavior consistent with the physical phenomena we can observe and duplicate in the laboratory. All our observations would suggest a common set of physical phenomena everywhere and at all times in the universe, which implies that the causal relationships we observe now are the same ones that pertained in the past. Our picture of the world is consistent with the assumption of causality. No one in fact would ever assume otherwise, if only because causality presents us with the only basis for understanding our experience and the only basis on which a science can be constructed. Hume's purpose however was not to have us disregard our observations or abandon science but to have us think about its foundations and to realize that our knowledge of the world, even that which we have demonstrated to ourselves again and again, can never be beyond doubt. Hume's objection still stands as one of the fundamental epistemological limitations to what we can know and how we can know it.

At the end of the nineteenth century, the physicist's world was not only causal but strictly deterministic. The outcome of any physical phenomenon was thought to be uniquely determined by the complete set of circumstances that gave rise to it. For any given set of physical conditions, one and only one outcome was possible. The circumstances of the moment completely determined the future course of events, just as the present was the uniquely determined result of whatever circumstances existed in the past. From a complete description of motion, both the future state of the universe and all of its past history could be re-created from knowledge of its present condition, at least in principle if not actually in practice.

The quantum description of motion has forced us to alter such a satisfying and straightforward view of the world. At the microcosmic level, a physical system exists simultaneously in all of its physically possible configurations, those allowed by the constraints of its physical parameters like mass and charge and the conservation of energy. Whenever we perform a measurement on the system, we find it in only one of its possible states, but which one is not uniquely determined by the conditions existing prior to the measurement. We cannot know before we make the measurement what the actual configuration will be. The measurement itself is involved in determining the configuration. Identical physical systems, configured and prepared and measured in exactly the same fashion, may yield different results among all of the possible outcomes. The wavefunctions that describe motion at the quantum level represent probability amplitudes that give us nothing more definite than the likelihood of finding a particle at a given location or with a particular momentum. Yet all of the information about the state of the system that we can have is contained in the wavefunction. The wavefunction of the total system is composed of all of the individual wavefunctions corresponding to each of the allowed physical configurations of the system. The interaction of the system with the measuring apparatus—its interaction with its surroundings—is required to determine which of the allowed configurations is actually observed; and the outcome is never certain but can be specified only in terms of a probability.

When a beam of photons impinges one at a time on a double slit, the pattern of transmitted light is the same as that produced by each photon going through both slits simultaneously. If we measure which slit each photon actually traverses, then the pattern changes to that produced by a single slit. In either case, the path of the transmitted photon cannot be predicted with certainty but can only be specified in terms of a relative probability. The same experiment can be performed using a beam of electrons instead of a beam of photons with identical results. Matter and energy consist of discrete units, or particles for lack of a better image, but these particles can simultaneously exist in multiple states and at multiple positions in space. We never observe matter and energy except in discrete quanta, but between measurements these fundamental units are not localized in the same way that our concept of a macroscopic particle would lead us to expect. The Schrodinger or the Dirac equation describes how these quanta move—and how the configuration of physical systems evolves—between measurements. The result is a description that is not strictly deterministic, in which the outcome of events cannot be predicted with certainty. Here again it is the description of motion, and not the measurement of the physical parameters by which we describe it, that is absolute and unequivocal. The wave equation is the same for all observers, but it can allow more than one outcome for any physical measurement.

The indeterminism of the material world is expressed by the uncertainty principle: the product of the uncertainty in certain pairs of variables like position and momentum, or energy and time has to be always greater than Planck's constant. We went for so long unaware of this indeterminism only because Planck's constant is such an exceedingly small quantity:

1.06×10^{-34} in our customary set of units. There is no reason to believe however that the description of motion at the quantum level does not apply equally to the macroscopic world. In fact, the description provided by Schrodinger's equation reduces in every case to that of Newtonian mechanics when we are dealing with quantities of matter and energy that are large compared to a quantum unit, a result referred to as the *correspondence principle*. What we measure in our observations of macroscopic events is the average behavior of large numbers of quanta, and the average behavior is more predictable than the behavior of a single quantum of matter or energy. The larger the number of quanta involved, the more precisely determined the average behavior of the ensemble becomes, until in the macroscopic realm the fluctuations due to an individual quantum are completely unobservable and lie outside the measurement capabilities of our instruments. On average, then, the world appears deterministic, but it is still made up of large numbers of indeterministic events between individual quanta. Our experience of this average behavior leaves us unacquainted with the underlying indeterminism, but it is there nevertheless.

Our knowledge of the world—indicated always by what we can measure—is limited at both extremes: by the finite speed of light in case of the very large and remote and by the indeterminism expressed by the uncertainty principle and Planck's constant in case of the very small and inaccessible. Both extremes are forever beyond our reach. The only real connection we can have with either is through the assumption of causality; and there we are limited by our inability to demonstrate that causality constitutes a necessary and essential condition of the physical world.

As shocking as these limits may seem at first to our deterministic view of the world, they should perhaps not be surprising. After all our understanding is based on our direct experience, plus the cumulative experiences that lie behind our evolution as a sensory and intelligent being, and that are now built into our way of interacting with the world. We do not form immediate judgments about what we are unable to perceive, except as inferences based on—but removed from—our direct experiences. Nor should we expect our mental processes to reflect anything beyond the material phenomena that gave rise to our evolutionary development. It seems only reasonable that, as creatures of sensory perceptions, we would be limited to what we can or have directly experienced, either during our lives or as a result of our evolution. Probably the most compelling evidence of causality is that we seem incapable of organizing and making sense of the world outside of this conceptual framework for interpreting our experience. That assumption was one of the cornerstones of Kant's argument in support of causality as an a priori condition of experience.

Due to our size, we have never directly experienced phenomena on an atomic or molecular scale or as individual atomic and molecular events. Nor can we have any direct experiences on the scale of light years or at speeds anywhere close to the speed of light. If we had been the size of atoms, we would not have conceived of Newtonian physics except as an approximate description applicable only to slow-moving massive objects composed of large quantities of matter and energy. In its place we would have arrived directly at the quantum theory as the proper description of physical reality. If we had been the size of galaxies, then the general theory of relativity would have seemed the correct description of the world and Newtonian physics would strike us as a crude approximation suitable only for small slow-moving objects in regions devoid of any large concentrations of matter and energy. Instead, we are almost midway between these two extremes, where the Newtonian description of motion adequately describes our experiences and we are at first unaware of the physical limitations imposed by either the speed of light or Planck's constant. For our local perceptions of the world the speed of light is large enough that it is essentially infinite, and Planck's constant is small enough that it might as well be zero. It was only when we achieved the necessary instrumental sophistication to push the precision of our measurements to the extremes that we became aware of either of these limits, and that became possible only in the twentieth century. Ironically, even though the limits have nothing to do with technology but are a fundamental feature of the world, it was technology—the technology involved in making more and more exacting measurements—that made it possible for us to discover these limits to our knowledge.

For most of the twentieth century, it was possible to reassure ourselves that the world of our sensory perceptions could still be adequately described by the laws of classical physics, represented by Maxwell's equations and Newton's laws of motion. Such a view of the world is tidy and reassuring. In it, macroscopic phenomena have only a single deterministic outcome and the motion of material bodies that are slow compared with the speed of light—which is always the case in our limited sensory experience—can be predicted with an accuracy far better than any achieved by our measuring instruments. Classical Newtonian physics describes the world better than our ability to measure what happens, which puts us in the reassuring position of having our ultimate knowledge subject only to practical considerations. But now we have discovered that at least some macroscopic phenomena are also inherently unpredictable even though they are deterministic. Certain nonlinear phenomena can, under the right circumstances, behave chaotically. When that happens, two sets of conditions, too close together to be distinguishable because of the limitations imposed by measurement or by the uncertainty principle, can lead to very different outcomes. In that case, our ability to predict the outcome fails even though the phenomenon itself is described by a completely deterministic equation.

This discovery of unpredictability delivered the final blow to the kind of simple determinism represented by classical physics. The limits to our knowledge now span the entire range of physical phenomena. We cannot know what happens outside of our immediate surroundings due to the finite value of the speed of light. Our knowledge about the interior of matter is limited by Planck's constant and the indeterminacy of quantum theory. And, in between, our ability to know the outcome of deterministic processes is limited by the chaotic behavior of nonlinear phenomena.

All processes occurring in nature involving interactions of mass and charge are inherently nonlinear. By that, we mean that the equations describing the motion of mass and charge are nonlinear equations, by virtue of the form taken by the mathematical expressions for the gravitational and electromagnetic forces. For sufficiently small excursions away from static equilibrium, we may be able to get away with describing the motion by linear equations, but for any finite motion the correct description is nonlinear. That does not mean, however, that all natural processes are chaotic. Just because the motion of a physical system is nonlinear does not imply that it is also chaotic. It does mean that it could have the potential for displaying the kind of sensitive dependence on initial conditions that would cause two virtually identical configurations of the system to lead to very different outcomes. It might also mean that under the correct set of circumstances—generally by being pushed far enough away from equilibrium—the resulting behavior of the system might become chaotic. The complete range of phenomena and the conditions under which natural systems can display chaotic behavior, along with the full range of behaviors that they can exhibit, has only begun to be investigated. We are still very much in the infancy of understanding nonlinear phenomena in nature. Under the proper conditions, nonlinear phenomena can exhibit stable, non-chaotic behavior corresponding to the remarkable stability that we observe in the material world around us. In most cases, variations in the initial configuration of a physical system do not become magnified but lead to only small differences in its final behavior, and the system remains both deterministic and predictable.

Except for situations that can be contrived in the laboratory, it may turn out that truly chaotic behavior in nature is relatively uncommon and has only a minor impact on the overall stability of the natural world. If our understanding of the universe is at all correct, there are processes in nature that continue to operate more or less the same for periods of time long even on a geologic or evolutionary scale. Stellar processes, the motion of the solar system, the development and evolution of life forms are obvious examples. It may also be that such seemingly stable processes are really the result of chaotic nonlinear systems that can make abrupt and unpredictable changes in behavior. There are systems in nature that can behave that way, and based on the geologic record the weather is one of them. We now know that global changes in the earth's climate such as the onset and the end of ice ages can occur rapidly even on a human time scale, and for reasons that are as yet poorly understood. Long-range weather forecasting appears to be impossible due to the sensitive dependence of the weather on the initial conditions of pressure, temperature, energy input and other parameters of the atmosphere, some of which we probably do not even know yet and certainly do not fully understand. The predictions can be no better than the mathematical description, and in this case even the crudest models are already nonlinear and describe behavior that is inherently unpredictable over any extended period of time.

Much of the present concern about global warming stems from just this situation. It isn't that anyone can say with certainty what effect continuing to increase the level of carbon dioxide in the earth's atmosphere will have on global climate, assuming that the measured increase is due to human activity, as seems likely. The concern is that we are dealing with a nonlinear system, one that if pushed sufficiently far from its current equilibrium may abruptly exhibit some totally unanticipated behavior.

Beyond that, one would be about as safe in predicting the onset of another ice age as in forecasting the melting of the polar ice caps. All anyone can say is that by our activity we may be increasing the risk of changes in the global patterns of weather that could disrupt human society. Such changes may occur anyway in spite of anything we could try to do about them. The problem with the remedies is that often they assume knowledge of how the system works and the ability to predict the outcome that those who propose them don't possess. The proposed remedies may be inconsequential, either because they are too little too late or because they have nothing to do with what actually causes global climate to behave as it does. When dealing with potentially chaotic systems the only sure way to find out what will happen is to wait and see, but such an attitude incorporates an element of risk and acknowledges a lack of control that we are reluctant to admit.

This brings us to the core concept of the book: *the enduring legacy of physics in the twentieth century has been the discovery of limits to our knowledge of the universe*. This message has been consistent and unequivocal and has been driven home repeatedly through each of the major developments during the past century: the theory of relativity, the quantum theory, and the discovery of deterministic chaos. Our knowledge and our understanding of the physical universe have increased enormously, and probably the most fundamental part of that increase has been the discovery of how our knowledge is limited by the nature of the world itself. For the first time since the beginnings of modern science in the seventeenth century, we have arrived at mathematical descriptions of nature that have built into them fundamental constraints on what it is *possible* for us to know. These are not limits that are likely to be supplanted or overcome by any future understanding. They appear to be inherent in the nature of motion itself and the concepts by which we measure and describe it. They are limits toward which the development of physical science has been pointed from the outset. Either our present understanding of the world is completely wrong—and given the operational success of our knowledge in investigating and manipulating our material surroundings that seems unlikely—or the limits we have discovered are indeed incontrovertible and inescapable. We should not be surprised to discover additional limits as we extend and refine our understanding in the future. There are unanswerable questions about the world; our awareness is a limited awareness. Any child can intuitively understand this before the age when our views of the world solidify into the set of beliefs and understandings by which we operate as adults. So too did the poets, long before Plato engaged them in dispute in the pages of the *Republic*. It is Plato and philosophy that bear the burden of proof. The twentieth century has vindicated the skepticism of the poets in declaring that the world at its core is unfathomable. We are born it seems with an intuitive understanding of our limitations, one that it has taken us more than twenty-five hundred years of concerted effort to finally verify in our present description of nature.

Ironically a similar development has been taking place at the same time in mathematics, which Plato held up as the model for philosophy, inscribing over the entrance to his Academy, "Let no one ignorant of mathematics enter here." Early in the twentieth century, David Hilbert, addressing a mathematical congress, confidently declared what no competent mathematician in the audience would have disputed. In a talk devoted to the most important problems remaining to be solved in mathematics, there were, he said, no questions in mathematics that would not eventually be answered, no mathematical problem without a solution. It was merely a matter of finding the correct approach, the correct method of attack.

A short time later one of Hilbert's own students, Kurt Godel, demonstrated that there are indeed formally undecidable propositions in any mathematical system that is extensive enough to encompass even the arithmetic of the whole numbers and that is also internally consistent, that is, not self-contradictory in some fashion. In such consistent mathematical systems, there are propositions that can be neither proved nor disproved. The truth or falsehood of such propositions is simply unresolvable. Thus any self-consistent mathematical system, like number theory, is always incomplete; one can ask questions about the properties of its members that cannot be answered. This shocking result is known as Godel's incompleteness theorem. Even more devastating was Godel's demonstration that the internal consistency of such mathematical systems also cannot be established. Not only are there unresolvable questions in any consistent mathematical system, one of the questions that cannot be resolved is whether the system is consistent. Thus there is built into the nature of mathematics, as there is in the nature of the material world, limitations to what we can know about the object of inquiry represented by mathematics-based systems of knowledge.

The public perception of physical science in the twentieth century has been far from this notion of the limitations of knowledge. We do not generally think of it as the century of limits and epistemological humility. What we see is that our automobiles operate more efficiently and reliably, airplanes fly faster, rockets go higher, computers perform unheard of tasks at greater and greater speeds, television peers into every nook and cranny of the globe. We have stood on the moon and sent space probes to the planets in the farthest reaches of the solar system, cured diseases, transplanted organs, deciphered the genetic code on which life itself is based, and placed all of this knowledge at the fingertips of anyone with access to the Internet. Surely knowledge of the world that makes these feats possible cannot be based on fundamental limits to what we can know. In the past century, as in no other before it, we believe we have become limited only by our ability to imagine what we shall attempt next.

The distinction here is between technology and science. Just because we are limited in what we can know does not mean that we are immediately limited in what we can do. Of course there are eventual limits to what we can do, the same ones that pertain to what we can know. We cannot travel about the universe at speeds greater than the speed of light. Nor can we build atomic-scale devices in which both the position and momentum of the components have to be specified simultaneously. Nor can we predict the weather in increasing detail arbitrarily far into the future. These are all tasks in which limits are imposed by the nature of the universe itself. But within these constraints we have enormous latitude as measured by our present technology and what it is still possible to achieve before encountering these limits.

General relativity tells us that we are confined to the past in our knowledge of the universe, but that does not alter our concept or understanding of past, present and future time in our immediate local surroundings. Quantum theory depicts a world that at the level of individual atomic events is indeterminant and governed by chance, but we continue to behave as if the macroscopic world is both deterministic and causal. We cannot demonstrate the necessity of causality but our actions show that we are convinced of it anyway. It is only in the physical sciences that the epistemological limits of twentieth century physics have had any real impact on how we think about the world. These are fundamental discoveries and constitute an important addition to our philosophical understanding of reality, but their immediate significance lies primarily beyond the reach of our senses. As long as we are dealing with macroscopic systems not too far removed from equilibrium, things that we can reach out and see and touch, then the world remains deterministic and for all practical purposes is adequately described by the classical physics of the nineteenth century.

It is only the physicist's world that is indeterminant and unpredictable and confined to the past. Biology by contrast is still very much a deterministic science. The biologist deals with structures that though small are still macroscopic compared with an atom, structures for which we can produce visible images with our microscopes and cameras. The biological world is still one that we can largely hold in our hands and see and touch and manipulate. Biology is only just now at the stage reached by physics at the end of the nineteenth century. In the quest to gain knowledge and understand the nature of living organisms, the biologist has yet to come up against the kind of fundamental limits encountered in physics. Of course at some atomic and molecular level biology must reduce to physics, and the limitations inherent in the quantum theory and in the behavior of nonlinear systems must manifest themselves in the internal atomic and molecular motions that constitute the fundamental processes of biology and the organization of biological systems. In our understanding of biology to date, such effects and their corresponding limits are not yet evident. The biologist represents the last scientific determinist. Witness the drive to understand the structure of the genes and decipher the genetic code, the ultimate goal that of being able to manipulate our own makeup and alter the nature of who and what we are. Such a prospect represents the ultimate in our ability to shape the world since it would give us control over our own biological definition and destiny.

The project of the human genome and the revolutionary possibilities inherent in genetic engineering will make biology the science of the twenty-first century. What seems likely to emerge are the fundamental laws governing the organization and collective behavior of large complex systems capable of self-replication. These laws will presumably take the form of a description detailing how such systems operate and evolve, analogous to the laws of physics describing the motion of matter and energy. The key to their discovery will probably lie in answering how self-replicating systems could have arisen in the first place. It will be interesting to see what limitations governing the behavior of genes arise from this description and to discover the limits imposed on our ability to shape and control that behavior. Only then will we learn the extent to which biology is a deterministic science or one subject to the same kind of fundamental restrictions inherent in the physicist's depiction of the world. The operating assumption for now seems to be that genes are deterministic structures governed by the laws of physics that we already know about and subject only to the limitations expressed by the uncertainty principle. If the historical development of science is any guide it would seem more likely that we will discover entirely new and as yet unanticipated principles governing the collective behavior of self-replicating systems—perhaps as windows of local order arising out of the chaotic behavior of nonlinear systems far from equilibrium, as already suggested by some.

For all of the limitations it imposes on our knowledge of the universe and our understanding of reality, the physicist's world ranks as one of humankind's most impressive intellectual achievements. The most remarkable and significant measure of its success can be seen in its ability to uncover the limits to its own inquiry. About itself the physicist's world provides specific and precise answers to the twin questions of epistemology with which we began: what can we know and how can we know it? We are able to describe the motion of matter and energy using only the measured properties of space and time. We require nothing actually beyond the measurement of distance. Time we gauge as the intervals between successive or repetitive events in space, like the swinging of a pendulum. Even the properties of matter—quantities like mass and charge—that embody our empirical observations of how things move are determined operationally by measuring combinations of distance and time such as velocity and acceleration.

The remainder of the answer is spelled out in the limits to knowledge embodied in the speed of light, the uncertainty principle, and the unpredictability of nonlinear systems. The answer may not have been what we were expecting, and it may not always agree with our intuition about the world given to us by our senses, but it is clear and unequivocal nonetheless. The physicist's world serves as the model and the standard not only for all of science but for all of epistemology as well. It is inconceivable that any consideration of epistemology could ever again be separate from an understanding of twentieth-century physics.

How do we account for this remarkable success? The answer is simple enough: physics succeeds so well because it attempts so little. All that it sets out to do is describe how material objects move. It does not attempt to explain the motion, only to describe it; or rather the only explanation it seeks to offer for why things move resides in the description of how they move in space and time. The physicist undertakes to answer only those *how* questions deemed beforehand to have answers; those intrinsically unanswerable questions of *why* are better left to poetry and philosophy. Even so we discover that some of the questions assumed at the outset to have answers do not, or at best have only partial answers. There are definite limits to how completely we can answer the question of how things move; or, equivalently, to how completely we can hope to understand the nature of the material universe. These are limits that we could not have known about before attempting the description, because they result not from the nature of the description itself but from the nature of the thing being described. We thought we were being modest in our goals and expectations, only to be rebuked by nature for having been too ambitious and presumptuous.

Everything else in physics—material properties like mass and charge, the structure of matter and the periodic table of the elements, the laws of chemistry governing the combination of elements to form chemical compounds, the types and properties of elementary particles making up the universe, the big bang model of cosmology—all of these are part of the description of motion. Each is in some way required to make sense out of the rest of the description. Physics does not give us any particular insight into the ultimate nature of mass or charge or any of the other intrinsic properties of elementary particles like leptons and quarks, beyond the set of measurements that we can perform to attach a numerical value to each of these properties. The physicist has no more revealing insight into the reality behind mass or charge than does the philosopher or the poet—only one that is precisely defined and rigidly confined by the description of motion that gives it meaning and the set of operations by which we measure them.

The success of physics in carrying out its program is not to be found in its means or its methods but in the modesty of its aspirations and goals. The physicist sets out to answer a simple question and to answer it in a straightforward manner. It is the simplicity of the question and not some characteristic of the approach or the methods that determines the success of the answer. The physicist uses mathematics not because there is some mysterious advantage in equations and formulas, but because mathematics is the most appropriate tool for dealing with quantitative concepts. Ever since the publication of the *Principia* in the seventeenth century, other disciplines have thought to emulate the remarkable success of physics by adopting its methods. Such efforts are largely misguided. There is little similarity between describing the trajectory of one particle moving under the influence of another, and trying to describe the various interactions between the many components of a complex social organization, other than perhaps the fact that both descriptions might use mathematics. The success of physics and its methods provides no insight at all into the much more complicated and intractable realm of human affairs, which belongs now, as it did in the time of Plato, to the poets and the philosophers.

It is a conceit of the present moment, if not much of the previous century, that we stand on the threshold of completely understanding the universe. Physicists speak glibly of a theory of everything and the eventual end of science. A theory of everything would presumably combine all of physics into one grand unified, and unifying, scheme that would explain the structure and workings of the entire universe, encompassing what is contained in all of the existing theories while filling in the missing pieces. The more euphoric of these prophets of science speak of a theory of everything as if it were a fait accompli, as if each of the necessary pieces were already sitting there just waiting for someone to assemble them in the correct manner. Even the more cautious and conservative of these optimists express little doubt that the task is well within our grasp, although no one among them has offered anything more substantial than the most embryonic suggestions about how, or even why, they believe such a scheme should be possible. One cannot help but be reminded with some amusement of Newton's unwavering belief in absolute space and time without being able to say what they were, to which in a sense all of the present euphoria can be traced.

The idea of a theory of everything is by no means new; nor is this the first time someone has prophesied the end of science. The ink was hardly dry on the general theory of relativity before Einstein and others were hard at work trying to combine gravity and electromagnetism into what became known as the unified field theory, one that would make gravity and electromagnetism merely different aspects of the same universal force, a task still uncompleted. Buoyed by the triumph of Maxwell's equations in describing electromagnetism and explaining the nature of light, physicists at the end of the nineteenth century publicly bemoaned that there was nothing left for them to do except measure more carefully the properties of matter and the fundamental constants of nature. Since that time the theory of relativity, the quantum theory, deterministic chaos, and two entirely new forces of nature, the nuclear strong force and the weak nuclear force, have been discovered and described. Following the publication of the *Principia*, Newton was hailed by his contemporaries as having discerned the mind of the Creator and explained the workings of the universe. What was meant, of course, was that he could describe, and in that limited sense only, explain, the motion of the heavenly bodies; yet Newton knew nothing of electromagnetism or the internal motion and composition of matter. It is almost axiomatic that, following every great discovery in physics, there are those for whom the final theory seems at hand and the end in sight.

To such absurdly outrageous claims we have only to compare the modest accomplishments chronicled in our story of the physicist's world. True, we have told only part of the story, but the part we have included is where matters are in reasonably good shape and we can be fairly confident of the results. In the portions we have omitted things are murkier and more speculative, even to the point of conceivably forcing us to have to revise our entire view of the physical world. And it is precisely from such speculations—in particular those dealing with particle physics and quantum gravity—that talk of a theory of everything emerges. It would seem a safer bet, and one more credible historically, to assume that rather than being at the end of anything we are on the verge of something entirely new and unexpected.

The Presocratic philosophers began with no less an ambition than that of understanding the nature of the universe. For all its notable success, we have had to settle for a much more modest description of how things move. The physicist's world is not something real and substantial. It is instead a creation of the human intellect and imagination. It is above all not synonymous with the tangible world of our senses, the world that we can reach out and touch and taste and that we can see and hear and smell. Least of all is it some grand and glorious tapestry on which the secrets of the universe are writ large for all to behold in wonder. It is a thinly woven web of mathematical equations by which we can describe the ebb and flow of matter and energy. The physicist's world is made from cloth exceedingly light and delicate, with but few strands in the warp and fewer still in the woof. Physics succeeds most where the description is the simplest, where the fewest mathematical equations describe the greatest diversity of phenomena, where the strands in its tapestry are as sparingly woven as human ingenuity can devise. And there are gaps in the weaving, not through any fault of the weaver but by design, natural—and human. The twentieth century has shown us that the blank spaces are an integral part of this tapestry. The description is limited in ways that are determined by the very nature of the thing being described. The omissions are as much a part of the universe as the silences between words or the blank spaces between sentences. Draped in the finest cloth that physics can weave, the emperor would still be very naked indeed.

Suggested Readings

These suggested readings are listed in roughly the order that follows the natural progression of the story told in *The Physicist's World*. They may of course be read in any order depending on the interests of the reader.

The Presocratics, Philip Wheelwright, The Odyssey Press, Bobbs-Merrill Publishing, 1960. All of the fragments of the Presocratic philosophers included in chapters 2–4 are taken directly from the translations given in Wheelwright. This work also contains commentary by later thinkers such as Plato and Aristotle on the writings and teachings of the Presocratic philosophers, along with introductory comments by Wheelwright on the nature of the problems and concerns addressed by the Presocratics. This book forms an excellent introduction to these thinkers as well as providing a complete collection of the surviving fragments of their writings.

The Nature of Things, Lucretius, translated by Frank O. Copley, W.W. Norton, 1977. An engaging and enjoyable translation of Lucretius's narrative poem into unrhymed loose iambic pentameter verse, in which the translator has taken special pains to convey the remarkable explanations of physical phenomena clearly and correctly. Copley's introduction provides valuable background information and insights into Lucretius's didactic poem.

The Way Things Are, Lucretius, translated by Rolfe Humphries, Indiana University Press, 1968. Another excellent rendering of Lucretius that is both readable and enjoyable. By comparing these two translations one can often get a clearer sense of Lucretius.

Five Dialogues, Plato, translated by G.M.A. Grube, Hackett, 1981. Contains translations and commentaries of the Socratic dialogues *Euthyphro*, *Apology*,

Crito, *Meno*, and *Phaedo* and forms an excellent introduction to the figure of Socrates in Plato's philosophy.

Plato, The Man and His Work, A.E. Taylor, Methuen, 1960. This has become one of the standard works on Plato and his philosophy.

Republic, Plato, the complete and unabridged Jowett translation, Vintage Books, 1991. Chapter X of *Republic* gives Plato's account of the "ancient quarrel" between poetry and philosophy and his arguments against the poets in defense of philosophy and a view of the world based on reason.

An Introduction to Ancient Philosophy, A.H. Armstrong, Rowman and Allenheld, 1983. Discussions of the philosophy of both Plato and Aristotle.

The Copernican Revolution, Thomas S. Kuhn, Harvard University Press, 1957. This thoroughly readable account provides a discussion of the extent to which the Ptolemaic world view and Aristotle's physics had become intertwined in the understanding of the physical world that existed during the middle ages and up to the time of the Copernican revolution at the beginning of the seventeenth century.

Siderius Nuncius or The Sidereal Messenger, Galileo Galilei, translated by Albert Van Helden, University of Chicago Press, 1989. In this little volume written for a general audience of his day, Galileo presented an account of his observations of the heavens using a telescope in support of the heliocentric model of Copernicus.

Dialogues Concerning Two New Sciences, Galileo Galilei, translated by Henry Crew and Alfonso de Salvio, Dover Publications, 1954. Galileo's last work, finished during his house arrest and summarizing his study of the problem of motion. Galileo's understanding of motion became the point of departure for Newton and later thinkers.

The Crime of Galileo, Giorgio de Santillana, University of Chicago Press, 1955. A complete and authoritative account of the conflict between Galileo and the Catholic Church, written by an eminent scholar who was given free access to church archives.

An Enquiry Concerning Human Understanding, David Hume, edited by Anthony Flew, Open Court Publishing, 1988. Hume's argument against the logical necessity of causality is presented in this work. The work also [illegible] tended argument against metaphysics and in support of sens[illegible] as the only basis of true knowledge.

Prolegomena to Any Future Metaphysics, Immanuel Kant, introductic[illegible] White Beck, Bobbs-Merrill Publishing, 1950. In this work Kant [illegible] to simplify and summarize his earlier response in the *Critique of Pure Reason* to Hume's argument against the logical necessity of causality. In it he outlines his argument for the possibility of a metaphysics based on synthetic a priori knowledge.

A Tour of the Calculus, David Berlinski, Pantheon Books, 1995. An informal intellectual history of the calculus and its development written for a wide audience and containing excellent expositions of the fundamental mathematical ideas of the calculus and the concepts of infinitesimals and continuity.

Newton's Philosophy of Nature: Selections from His Writings, edited by H.S. Thayer, Hafner Press, Macmillan, 1953. These selections make Newton accessible to the modern reader unequipped to delve directly into the *Principia* and include important sections from that work plus from Newton's letters and other writings. All of the quotations of Newton in the text are taken directly from this work.

Foundations of Physics, Robert Bruce Lindsay and Henry Margenau, Dover Publications, 1957. This standard work on the mathematical and philosophical foundations of physics contains an excellent, though at points mathematical, discussion of Newton's laws of motion along with a treatment of the general problem of motion including the theory of relativity.

The Concept of Energy Simply Explained, Morton Mott-Smith, Dover Publications, 1964. The principles of thermodynamics including the first and second laws are presented and developed in their historical and technical context. An excellent and accessible introduction for the lay reader to the practical concepts of thermodynamics.

Relativity, A Clear Explanation That Anyone Can Understand, Albert Einstein, Crown Publishers, 1961. Einstein's own explanation of both the special and general theories of relativity written immediately after he finished the theory of general relativity and intended by him for the lay reader. Even if you end up disagreeing with the subtitle of this little book it is still well worth reading. It provides valuable insights into the way that Einstein himself thought about and tried to convey the concepts of special and general relativity.

General Relativity, I.R. Kenyon, Oxford University Press. 1990. An undergraduate level text that presents an introduction to the complete theory of general relativity at probably the most elementary (though still difficult) level consistent with its mathematical nature.

Wholeness and the Implicate Order, David Bohm, Routledge, 1980. In this book Bohm explores some of his ideas about alternative interpretations of quantum mechanics as opposed to the standard Copenhagen interpretation.

QED, The Strange Theory of Light and Matter, Richard P. Feynman, Princeton University Press, 1985. One of the best and most intellectually honest of the popular expositions of quantum electrodynamics, by one of its founders and a master at exposition.

Space, Time and Gravity, second edition, Robert M. Wald, University of Chicago Press, 1997. A summary for the nontechnical reader of current thinking about the theory of general relativity and its applications to cosmology.

Black Holes and Time Warps, Kip S. Thorne, W.W. Norton, 1994. A comprehensive and readable account of the phenomena of general relativity and modern cosmology.

Chaos, James Gleick, Penguin Books, 1987. Of the many popular treatments of deterministic chaos, this remains one of the most comprehensive and accessible accounts of the broad range of topics and phenomena included in this diverse and as yet not-well-defined field overlapping a number of disciplines.

Order Out of Chaos, Ilya Prigogine and Isabelle Stengers, Bantam Books, 1984. Although it sometimes reads like the work of a committee whose members did not talk to one another often enough, this work contains interesting insights and speculation about some of the possible implications of deterministic chaos for self-organizing systems.

Chaotic Dynamics, An Introduction, second edition, G.L. Baker and J.P. Gollub, Cambridge University Press, 1996. An introductory level, though mathematical, undergraduate text to the field of deterministic chaos applied to the description of physical phenomena.

The Elegant Universe, Brian Greene, Vintage, 1999, and *The Trouble with Physics*, Lee Smolin, Houghton-Mifflin, 2006. Two books on string theory written by theoretical physicists. The first is by a true believer who sees in the direction suggested by string theory the potential answer to all of physics' problems and the road to an eventual Theory of Everything. The second is by a vocal critic of string theory who sees it as a misguided agenda and an impediment to any real progress in particle physics and achieving a theory of quantum gravity. These two books should be read and enjoyed together.

Index

The av rate of change of f(x) on [a, b]
is f(b)-f(a)/ b-a. Gradient